Innovative Methods in Support of Bioremediation

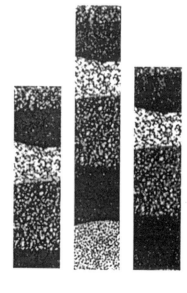

Editors

Victor S. Magar, Timothy M.
Vogel, C. Marjorie Aelion, and
Andrea Leeson

The Sixth International In Situ and
On-Site Bioremediation Symposium

San Diego, California, June 4–7, 2001

BATTELLE PRESS
Columbus • Richland

Library of Congress Cataloging-in-Publication Data

International In Situ and On-Site Bioremediation Symposium (6th : 2001 : San Diego, Calif.)
 Innovative methods in support of bioremediation : the Sixth International In Situ and On-Site Bioremediation Symposium : San Diego, California, June 4-7, 2001 / editors, V. Magar ... [et al.].
  p. cm. -- (The Sixth International In Situ and On-Site Bioremediation Symposium ; 4)
 Includes bibliographical references and index.
 ISBN 1-57477-112-4 (hc. : alk. paper)
  1. Bioremediation--Technological innovations--Congresses. 2. Bioremediation--Congresses. I. Magar, V. (Victor), 1964- . II. Title. III. Series: International In Situ and On-Site Bioremediation Symposium (6th : 2001 : San Diego, Calif.). Sixth International In Situ and On-Site Bioremediation Symposium ; 4.
 TD192.5.I56 2001 vol. 4
 628.5 s--dc21
 [628.5]
                                                                                    2001044131

Printed in the United States of America

Copyright © 2001 Battelle Memorial Institute. All rights reserved. This document, or parts thereof, may not be reproduced in any form without the written permission of Battelle Memorial Institute.

Battelle Press
505 King Avenue
Columbus, Ohio 43201, USA
614-424-6393 or 1-800-451-3543
Fax: 1-614-424-3819
Internet: press@battelle.org
Website: www.battelle.org/bookstore

For information on future environmental conferences, write to:
 Battelle
 Environmental Restoration Department, Room 10-123B
 505 King Avenue
 Columbus, Ohio 43201-2693
 Phone: 614-424-7604
 Fax: 614-424-3667
 Website: www.battelle.org/conferences

CONTENTS

Foreword   v

Risk Assessment, Toxicity, and Treatment Costs

Ecological Risk Assessment in Derwent Estuary, Tasmania: A Critical Evaluation.
Z. San Felipe, R. Beckett, and B. Hart   1

Toxicity Testing of Weathered Biotreated Petroleum Hydrocarbon-Contaminated Soils. S. Murphy and J. Charrois, and W. McGill   7

A Method for Assessing the Full Costs and Benefits of Groundwater Remediation.
P.E. Hardisty, J. Dottridge, S. Wallace, J. Smith, E. Ozdemiroglu, and J. Fisher   17

Molecular Monitoring Techniques and Microbial Enumeration

Selective Molecular Probes for Detecting Aromatic-Compounds Degrading Bacteria in Environmental Samples. M. Martin, M. Sanchez, C. Garbi, E. Ferrer, A. Gibello, J.L. Allende, M.J. Martinez-Inigo, C. Lobo, and C. Martin   29

Monitoring Microbial Community Changes in Oil-Contaminated Laboratory Soil Microcosms. K. VanBroekhoven, R. De Mot, L. Bastiaens, J. Gemoets, and D. Springael   35

Field Release of Genetically Engineered Bioluminescent Bioreporters for Bioremediation Process Monitoring and Control. S. Ripp, D.E. Nivens, and G.S. Sayler   45

Rapid, Sensitive, and Accurate Monitoring Method for Augmented Dioxin-Degrading Bacteria. H. Nojiri, J. Widada, T. Yoshida, H. Habe, and T. Omori   51

Monitoring Bioremediation Through In Situ Soil Respiration. C.C. Nestler, L.D. Hansen, S. Waisner, and J.W. Talley   59

Assessment of Alternative Endpoints in Landfarming Systems for Sustainable Soil Use. K.C. Nieman, R.C. Sims, and H.-Y. N. Holman   67

Microbial Heterogeneity Implications for Bioremediation. S.M. Pfiffner, A.V. Palumbo, B.L. Kinsall, A.D. Peacock, D.C. White, and T.J. Phelps   73

Degradation of Pyridine by *Rhodococcus opacus* UFZ B408. C.G. Follner and W. Babel   81

Carbon Isotope Analyses for Monitoring Biotransformation Processes

Natural Isotope Analysis: A Promising Tool in Soil Pollution Research.
F. Volkering, H. Jonker, B.M. van Breukelen, J. Groen, H.A.J. Meijer,
B. Sherwood-Lollar, and J.D. Kramers   91

Characterisation of Microbial In Situ Degradation of Aromatic Hydrocarbons.
H.H. Richnow, M. Gehre, M. Kastner, B. Morasch, and R.U. Meckenstock   99

In Situ Vinyl Chloride Biodegradation Revealed Through Carbon Isotope
Composition. D. Graves, G.R. Hecox, K. Kirschenmann, S. Ingram, and
B. Sherwood Lollar   109

One Year of Monitoring Natural Attenuation in Shallow Groundwater.
L.G. Stehmeier, R. Hornett, L. Cooke, and M.McD. Francis   117

Modeling

Model Analysis of Reductive Dechlorination with Data From Cape Canaveral
Field Site. C. Shoemaker, M. Willis, W. Zhang, and J. Gossett   125

Bioremediation Modeling: From the Pilot Plant to the Field. M. Villani,
M. Padovani, M. Andretta, M. Mazzanti, R. Serra, B. Mueller, H.P. Ratzke, R. Rongo,
W. Spataro, and S. Di Gregorio   131

Modeling the Dynamics of Biodegradation in Unsaturated Soil for
Implementation of Advanced Control Strategies. O. Schoefs, R.P. Chapuis,
M. Perrier, and R. Samson   139

Comparision of Experimental and Simulated Results of a Rock-Bed Filtration
Model. R. Jindal and S. Fujii   149

Author Index   159

Keyword Index   187

FOREWORD

The papers in this volume correspond to presentations made at the Sixth International In Situ and On-Site Bioremediation Symposium (San Diego, California, June 4-7 2001). The program included approximately 600 presentations in 50 sessions on a variety of bioremediation and supporting technologies used for a wide range of contaminants.
This volume focuses on *Innovative Methods in Support of Bioremediation*. New monitoring techniques aimed at characterizing microbial consortia and monitoring microbial activity at hazardous waste sites continue to gain increasing use in the industry, and several such methods supported by field applications are presented. The use of isotopes to confirm in situ biological transformation of chlorinated compounds and fuel hydrocarbons at enhanced bioremedation and natural attenuation sites is demonstrated in several papers. Novel modeling applications to support the design of enhanced bioremediation methods and to verify natural attenuation are presented.
The author of each presentation accepted for the symposium program was invited to prepare an eight-page paper. According to its topic, each paper received was tentatively assigned to one of ten volumes and subsequently was reviewed by the editors of that volume and by the Symposium chairs. We appreciate the significant commitment of time by the volume editors, each of whom reviewed as many as 40 papers. The result of the review was that 352 papers were accepted for publication and assembled into the following ten volumes:

Bioremediation of MTBE, Alcohols, and Ethers — 6(1). Eds: Victor S. Magar, James T. Gibbs, Kirk T. O'Reilly, Michael R. Hyman, and Andrea Leeson.
Natural Attenuation of Environmental Contaminants — 6(2). Eds: Andrea Leeson, Mark E. Kelley, Hanadi S. Rifai, and Victor S. Magar.
Bioremediation of Energetics, Phenolics, and Polycyclic Aromatic Hydrocarbons — 6(3). Eds: Victor S. Magar, Glenn Johnson, Say Kee Ong, and Andrea Leeson.
Innovative Methods in Support of Bioremediation — 6(4). Eds: Victor S. Magar, Timothy M. Vogel, C. Marjorie Aelion, and Andrea Leeson.
Phytoremediation, Wetlands, and Sediments — 6(5). Eds: Andrea Leeson, Eric A. Foote, M. Katherine Banks, and Victor S. Magar.
Ex Situ Biological Treatment Technologies — 6(6). Eds: Victor S. Magar, F. Michael von Fahnestock, and Andrea Leeson.
Anaerobic Degradation of Chlorinated Solvents— 6(7). Eds: Victor S. Magar, Donna E. Fennell, Jeffrey J. Morse, Bruce C. Alleman, and Andrea Leeson.

Bioaugmentation, Biobarriers, and Biogeochemistry — 6(8). Eds: Andrea Leeson, Bruce C. Alleman, Pedro J. Alvarez, and Victor S. Magar.
Bioremediation of Inorganic Compounds — 6(9). Eds: Andrea Leeson, Brent M. Peyton, Jeffrey L. Means, and Victor S. Magar.
In Situ Aeration and Aerobic Remediation — 6(10). Eds: Andrea Leeson, Paul C. Johnson, Robert E. Hinchee, Lewis Semprini, and Victor S. Magar.

In addition to the volume editors, we would like to thank the Battelle staff who assembled the ten volumes and prepared them for printing: Lori Helsel, Carol Young, Loretta Bahn, Regina Lynch, and Gina Melaragno. Joseph Sheldrick, manager of Battelle Press, provided valuable production-planning advice and coordinated with the printer; he and Gar Dingess designed the covers.

The Bioremediation Symposium is sponsored and organized by Battelle Memorial Institute, with the assistance of a number of environmental remediation organizations. In 2001, the following co-sponsors made financial contributions toward the Symposium:
Geomatrix Consultants, Inc.
The IT Group, Inc.
Parsons
Regenesis
U.S. Air Force Center for Environmental Excellence (AFCEE)
U.S. Naval Facilities Engineering Command (NAVFAC)
Additional participating organizations assisted with distribution of information about the Symposium:
Ajou University, College of Engineering
American Petroleum Institute
Asian Institute of Technology
National Center for Integrated Bioremediation Research & Development (University of Michigan)
U.S. Air Force Research Laboratory, Air Expeditionary Forces Technologies Division
U.S. Environmental Protection Agency
Western Region Hazardous Substance Research Center (Stanford University and Oregon State University)

Although the technical review provided guidance to the authors to help clarify their presentations, the materials in these volumes ultimately represent the authors' results and interpretations. The support provided to the Symposium by Battelle, the co-sponsors, and the participating organizations should not be construed as their endorsement of the content of these volumes.

Andrea Leeson & Victor Magar, Battelle
2001 Bioremediation Symposium Co-Chairs

ECOLOGICAL RISK ASSESSMENT IN DERWENT ESTUARY, TASMANIA: A CRITICAL EVALUATION

Zenaida San Felipe, Ronald Beckett, Barry Hart
(Monash University, Clayton, Victoria, Australia)

ABSTRACT: This paper presents the evaluation on Norske Skog Paper Mill Ecological Risk Assessment (NSPA-ERA) conducted in the Derwent Estuary. This study evaluates the NSPA-ERA approach by using the U.S. Environmental Protection Agency's framework for ecological risk assessment as a reference. Strong similarity exists between the NSPA-ERA approach and the approach recommended by the U.S. Environmental Protection Agency framework for ecological risk assessment (USEPA-ERA). However, there are some notable differences, which are presented and discussed.

INTRODUCTION

Paper mill effluents have the potential to cause changes to physico-chemical properties of aquatic environments. Areas affected by such changes are dependent on volumes and rates of effluent discharges. Specifically, effluents have the potential to affect aquatic biota. Biota could be affected either by chemical activity of wastes or via changes in physico-chemical environmental conditions. Discharged compounds are likely to be taken up by biota across respiratory surfaces, by ingestion, or through nutrient uptake processes, either directly from sediments or from the water column.

Impacted aquatic systems must be assessed to identify the factors responsible for degradation. Based on results of the assessment, remedial action/s to the problem could be implemented in most cost-effective and efficient manner.

Increasingly, ecological risk assessment is used as a tool to evaluate effluent impacts, within a specified range of uncertainty. Ecological risk assessment provides a sound scientific basis for determining appropriate limits for pollutant concentrations in whole ecosystems (Bennett & Twining 2000).

Norske Skog Paper Mills (NSPA) embarked on an ecological risk assessment, the main objective of which was to determine whether "environmental harm" to Derwent Estuary is being caused by NSPA's combined effluent stream (NSR 1999). This paper presents results of the evaluation on NSPA ecological risk assessment (NSPA-ERA) process.

Objective. The objective of this study was to evaluate the NSPA-ERA conducted for the Derwent Estuary. Specifically, the aim was to identify strengths and weaknesses of the NSPA-ERA. Recommendations are proposed in order to improve the NSPA-ERA process.

METHODOLOGY

The ERA conducted for the Derwent Estuary was evaluated by comparing the NSPA-ERA approach with the USEPA-ERA approach (USEPA, 1998). The evaluation focuses on processes employed in three stages of the ERA process (Stage 1 – Planning and Problem Formulation, Stage 2 – Analysis, and Stage 3 – Risk Characterization).

A literature search and review of risk assessment databases was conducted to achieve an overview of the development, scope, importance, and issues regarding ERAs. A site-visit to the paper mill was conducted in November, 1999. Processes employed by the NSPA-ERA was then compared with ERA processes enumerated in a checklist prepared by the researchers. The checklist consists of ERA processes recommended by the USEPA-ERA.

Based on results of the evaluation, recommendations are proposed to refine the site-specific NSPA-ERA and, more specifically, to improve the quantitative assessment of ecological effects of different stressors from the mill.

RESULTS AND DISCUSSION

Paper mills differ mostly in their feedstock and pulping process. Apparently, this is the main reason why it is not possible to generalize ERA results from other aquatic systems receiving paper mill discharge and apply to another site as the Derwent Estuary.

NSPA-ERA approach has many points of similarity with some current ERA approaches in Australia and especially with USEPA-ERA. The following are the similarities between the NSPA-ERA and the USEPA-ERA:

- Emphasis on efficient communication and coordination between risk assessors and management group.
- Ensure accurate and updated documentation of results of the ERA.
- Problems or gaps of information identified in any stage of the process direct risk assessors to go back to previous step/s in the ERA process.
- Relationship between stressors (chemicals in the mill effluent), assessment endpoints (ecological values), and exposure pathways are represented by conceptual models.
- Focus more on quantitative descriptions of risks.
- Conduct planning dialogue between risk assessors and management group.
- Scientific conduct of ERA (referred to several lines of evidence to characterize risks and provide quality assurance/quality control to ensure validity of data.

Stage 1- Screening Level Study (planning and problem formulation phase) of the NSPA-ERA identified contaminants of possible concern, screened out other contaminants, and defined data requirements. The planning and problem formulation phase helped focus the scope of ERA. That is, stressors were screened for whether they would need further assessment. Stressors that warranted further assessments were identified and carried to the Analysis phase or Stage 2.

Conservative assumptions were used in screening assessments to ensure that no contaminants were inappropriately excluded. Conservative assumptions in screening stressors are generally consistent with USEPA-ERA. By doing so, the final estimate of risk will be higher than the upper end of likely range of actual risks (USEPA 1989, USEPA 1993).

NSPA-ERA used two approaches to screen various stressors: (1) comparison to benchmark values and (2) comparison with background concentrations. In benchmark comparisons, chemical concentrations detected in environmental media or in tissues of exposed organisms were compared with published criteria or guidelines to assess severity of contamination. The statistical descriptor chosen by NSPA-ERA is the 95% upper confidence limit of the mean (NSR 1999). Benchmark comparisons are also widely used in ERA conducted in the U.S., Canada, and Europe.

Conceptual models in NSPA-ERA process are developed based on information about stressors and potential exposure and predicted effects on assessment endpoints. NSPA-ERA conceptual models are used to represent stressors, assessment endpoints, responses, exposure routes, and ecosystem processes. In addition, hypotheses generated from conceptual models describe the approach to be used for the analysis stage and identify types of data and analytical tools that will be needed. USEPA-ERA also uses conceptual models in the assessment of environmental impacts caused by various stressors.

This study found the following differences between the NSPA-ERA and the USEPA-ERA:

- Ideally, reference sites are used in ERAs. Ecological risk assessments in the U.S. which use field surveys, bioassays, and toxicity tests to assess biological changes compare measures of effect at the affected site and the same measures of effects at one or more comparable reference sites (Henning 1998).

While the use of reference sites is generally the preferred approach in ERA, it is not considered relevant within the context of NSPA-ERA (NSR 1999). This is because (1) there is no suitable historical unimpacted data; (2) although upstream sites are included in the proposed main channel characterization, nature of Derwent River changes substantially downstream from the mill limiting the applicability of data, particularly with respect to biological indicators; (3) concerning both local and upstream sites, conditions at such sites do not reflect high degree of historical disturbance evident downstream of the mill, that is, environmental characteristics between potential control sites and sites of interest are too dissimilar (NSR 1999).

In as much as no applicable site was identified that could serve as reference in the NSPA-ERA, modelling has proven to be a better alternative in the NSPA-ERA. Mathematical modelling was conducted by CSIRO Division of Marine Research. Three models of the upper Derwent Estuary were implemented and integrated by CSIRO Division of Marine Research. Predictions generated from the modelling have closely matched field observations.

- USEPA-ERA involves the community in the planning stage of ERA. In NSPA-ERA, however, community participation was initiated only at the later stage of the ERA. USEPA-ERA regards community participation as an integral factor in the planning phase of ecological risk assessment. The community can provide important information/local knowledge about the site to be assessed and guidance on priority issues that risk assessors need to consider in the assessment process. Community participation that is proactive and occurs early in the USEPA-ERA process as an input to the planning phase is seen to be preferable to reactive community participation that is focused on mitigation.

Further, community participation gives them (community) more opportunities for airing their interests and opinions about the ecosystem to be assessed. Their opinion is considered vital to the planning stage of the USEPA-ERA. Thus, it is recommended that NSPA-ERA could improve the planning stage by providing community participation. The community has the right to be informed/involved with any undertaking which could directly/indirectly affect them.

NSPA-ERA findings provide a summary information concerning the present status of the impact of effluent discharge to the Derwent Estuary. NSPA-ERA was successful in quantitatively describing risk uncertainties using specific assessment tools.

Both the NSPA-ERA and the USEPA-ERA help ensure that management actions remain relevant and help sustain the environment. Thus, a modified/improved NSPA-ERA can be used as a model for the development of an integrated ERA approach for cartchment-wide protection and management in Australia.

ACKNOWLEDGEMENT

This paper would not have been possible without the help and cooperation of Norske Skog Paper Mills (Australia) Ltd. Thanks are also due to Water Studies Center at Monash University, CRC for Freshwater Ecology, NSR Environmental Consultants Pty Ltd., Victoria EPA. The first author also wishes to thank for the funds provided through her scholarship in the Australian Assistance for International Development (AUSAID).

REFERENCES

ANZECC (1992). Australian and New Zealand Environment and Conservation Council. *Australian Water Quality Guidelines for Fresh and Marine Waters*. ANZECC, Canberra.

Bennett, J.W. and J.R. Twining (2000). *Australian Minerals and Energy Environment Foundation*. Yearbook. Best Practice 2000.

Henning, M. and N. Shear. 1998. Regulatory Perspectives on the Significance of Ecological Changes as Reported in Ecological Risk Assessment. *Human and Ecological Risk Assessment.* 4 (4): 807-814.

NSR. 1999. *Ecological Risk Assessment. Stage 1 – Screening Level Study.* NSR Environmental Consultants Pty. Ltd.

USEPA. 1989. *Risk Assessment Guidance for Superfund: Human Health Evaluation Manual.* Part A Interim Final. Office of Solid Waste and Emergency Response. OSWER Directive 9285.701 A.

USEPA, 1992. *Framework for Ecological Risk Assessment. Risk Assessment Forum.* U.S. Environmental Protection Agency. EPA/630/R-92/001. Washington, D.C.

USEPA. 1993. Integrated Risk Information System (IRIS). *Chemical-Specific Reference Doses and Cancer Potency Factors and EPA Toxicology Background Documents.* Office of the Health and Environment Assessment Environmental Criteria and Assessment Office. Cincinnati, Ohio.

USEPA. 1998. *Guidelines for Ecological Risk Assessment.* Risk Assessment Forum. U.S. Environmental Protection Agency. EPA/630/R-95/002F. Washington, D.C.

TOXICITY TESTING OF WEATHERED BIOTREATED PETROLEUM HYDROCARBON CONTAMINATED SOILS

Sean Murphy, Jeffrey Charrois (Komex International Ltd., Edmonton, Alberta)
William McGill (University of Northern British Columbia,
Prince George, British Columbia)

ABSTRACT: The Canadian Council of Ministers of the Environment (CCME) recently developed standards for petroleum hydrocarbon (PHC) contamination in soil. These standards are based substantially on toxicity tests conducted with different carbon fractions of PHC freshly added to pristine soils. Risk assessment and continued refinement of standards require data relating PHC concentrations in aged samples to measured toxicity. A battery of laboratory acute and subchronic toxicity studies was initiated in the summer of 2000 to evaluate: 1) the toxicity of residual PHC in four weathered biotreated soils and a pristine control soil; and 2) the CCME PHC standards against the hazard of weathered hydrocarbons. Hydrocarbon contaminated soils were collected from four upstream oil and gas facilities in Alberta, Canada. All soils were biotreated until hydrocarbon degradation had reached a plateau and further reductions appeared unlikely. All soils were in excess of CCME PHC standards for various soil types, land uses and exposure pathways, because of the F3 PHC fraction. We measured no toxicity from the battery of toxicity tests used, i.e. Microtox™, seed germination (> 90 % germination), and earthworm survival (100 % survival to 10 weeks of exposure). We conclude that the risk of toxicity from weathered PHC in bioremediated soils may be overestimated by current CCME PHC standards.

INTRODUCTION

The Canadian Council of Ministers of the Environment (CCME) has recently implemented Canada-wide standards for petroleum hydrocarbons (PHCs) in soil (CCME, 2000). The CCME PHC standards designate four hydrocarbon fractions: 1) Fraction (F) 1:C_6-C_{10}, 2) F2:>C_{10}-C_{16}, 3) F3:>C_{16}-C_{34} and F4:>C_{34}-C_{50}. These fractions have their own associated criteria based upon the likelihood that constituents within a given fraction could produce a potential environmental or human health risk. The rationale being those hydrocarbon fractions that may pose a greater risk (*i.e.* C_6-C_{10} and >C_{10}-C_{16}) are more highly regulated. The CCME standards also take into account soil texture (*i.e.*, fine versus coarse), soil location (*i.e.*, surface versus subsoil), possible land uses (*i.e.*, agricultural, residential, commercial and industrial) and exposure pathways (*i.e.*, soil ingestion, vapor inhalation and eco-soil contact *etc.*). Hydrocarbon contaminated soils were collected from four upstream oil and gas facilities in Alberta, Canada. All soils were biotreated until hydrocarbon degradation had reached a plateau and further reductions appeared unlikely.

Bioremediation is widely used to treat hydrocarbon contaminated soils; however following treatment some residual contamination generally remains (Angehrn et al., 1998). Interpretation of hydrocarbon data (both concentrations and chromatograms) during bioremediation trials (data not presented) revealed that degradation had reached a plateau and that residual hydrocarbon contamination remained in all four treated soils. Hydrocarbon characterization revealed that PHC within all four soils generally had a carbon distribution between C_{15}-C_{40} and was a weathered, unresolved complex mixture (UCM). The biotreated material contained no detectable volatile hydrocarbon compounds (C_5-C_{10}) corresponding to the CCME PHC F1 fraction. Additionally, it is assumed that the majority of water-soluble organic compounds would have been degraded during treatment.

The reasons for the plateau in hydrocarbon concentrations during bioremediation are well documented in the literature and are often generalized as the 'hockey stick' effect (Alexander, 1995). The explanation for this observed phenomenon is related to a combination of: 1) sequestration of contaminants within inaccessible pore spaces and voids, 2) rate-limited mass transfer associated with both hydrophobic organic contaminant transfer to the aqueous phase and diffusion of more readily degradable compounds from deeper within the soil matrix to the aqueous phase; and 3) bioavailability issues surrounding the aforementioned physical-chemical limitations in conjunction with a lack of microbial access to amenable organic substrates.

The CCME PHC standards are based substantially on toxicity tests conducted with different carbon fractions of PHC freshly added to pristine soils. Toxicity of a xenobiotic in soil varies with its bioavailability. Bioavailability of freshly added PHC may not represent that of aged PHC. Recent research has demonstrated that organic compounds that have been aged in the field may have appreciably lower bioavailability than the same compounds freshly added to soil (Alexander, 2000). Risk assessment and continued refinement of standards require data relating PHC concentrations in aged samples to measured toxicity. CCME has indicated that there remains a known information gap related to the ecotoxicology of weathered PHC (CCME, 2000). A battery of acute and subchronic toxicity studies was initiated in the summer of 2000 to evaluate: 1) the toxicity of residual PHC in four weathered biotreated soils and a pristine control soil; and 2) the CCME PHC standards against the hazard of weathered hydrocarbons.

MATERIALS AND METHODS

Soils. Four PHC contaminated soils were collected from various upstream oil and gas facilities in Alberta, Canada. Soil samples were collected to reflect a range of hydrocarbon concentrations (relative to CCME guidelines) as well as to represent different sources of contamination. Samples were collected from land treatment areas, once a plateau in hydrocarbon degradation was observed in sample chromatograms and further reductions in PHC levels seemed unlikely.

The origin of contamination for Soils A, C, and D was from sump material collect at three different sites, while the source of contamination for Soil B was from a wellhead. In general sump material contains contamination that includes crude oil and drilling wastes, while wellhead contamination contains only crude oil. Appropriate control soils were not available at or near the contaminated sites so a pristine control soil was selected based on correlative soil variables. The main variables used for selection of the pristine soil were pH, electrical conductivity (EC), texture, and bulk density.

Physical and Chemical Characterization. After the soils were collected from each of the four land treatment areas, samples were taken to the lab and prepared for chemical and physical analyses. All collected samples were air dried at 50 °C for 18 hours ± 2 hours. Each sample was then ground to a maximum 2 mm aggregate size using a manual grinder. Samples from each site were combined into one large composite sample and stored in a 1-gallon pail. The composite samples were homogenized by stirring with a wooden stick and by rolling the pail on the ground for 10 minutes. Following homogenization, each site composite sample was then split into three subsamples. Analysis for particle size distribution within the subsamples was used to evaluate the relative homogeneity of each subsample prior to subsequent inclusion in characterization and/or experimental trials.

Soils were analyzed for PHC, polycyclic aromatic hydrocarbons (PAHs), pH, EC, and particle size. PHC concentrations were determined using protocols outlined by the CCME (CCME, 2000). Briefly, quantification of PHC fraction F1 involves a methanol extraction and purge and trap procedure followed by GC/FID. Fractions F2-F4 (up to C_{50}) are first extracted using a Soxhlet extraction (50:50 hexane:acetone) procedure, which is then dried using sodium sulfate and treated with silica gel (to remove polar materials). The extract is then analyzed using GC/FID analysis. For the C_{50+} fraction, either a gravimetric or extended high-temperature chromatography determination is made (CCME, 2000). For the purposes of this study, historical data (not presented) suggested that characterization of the PHC F1 (C_6-C_{10}) fraction was not necessary. The PHC F1 fraction is the most volatile and since these soils were biotreated in the field for several years, sufficient opportunity was provided for the volatile components to degrade/dissipate. Soil PAH analysis followed EPA 3545, EPA 8270, and EPA 3611B procedures. Soil chemical analyses were conducted in triplicate and values are reported as the average of replicate samples (result) ± 95 % confidence interval.

Toxicity Testing. Toxicity bioassays, conducted in triplicate, consisted of: Microtox™, radish (*Raphanus sativus*) and oat (*Avena sativa*) seed germination; as well as acute (14 day) and subchronic (10 week) exposures of the earthworm (*Eisenia fetida*).

The Microtox™ toxicity assay was conducted on 1:1 water extracts from the four biotreated soils according to procedures outlined by Environment Canada (Environment Canada, 1992). The microbial toxicity assay was conducted in triplicate for each contaminated soil.

A seed germination bioassay was conducted on each contaminated soil and the control soil. Radish (*Raphanus sativus*) and oats (*Avena sativa*) were selected for seed germination trials. The first exposure was to 100 % contaminated material, the second treatment was a 50 % dilution of the contaminated material with a pristine control soil (w/w), and the final treatment was a control soil. The germination assay followed the principles of ASTM method E 1598-94 and was done in triplicate.

Earthworm survival was the endpoint used for the toxicity assays described here. The earthworm toxicity bioassay incorporated many of the principles described in ASTM, (1995); and Greene et al (1989). Briefly, the protocol is outlined below. Mature earthworms (*Eisenia fetida*) were used for all acute (14 day) and subchronic (10 weeks) toxicity tests. Soil treatments for the acute and subchronic toxicity testing consisted of the three exposure treatments, conducted concurrently. The first exposure was to 100 % contaminated material, the second treatment was a 50 % dilution (data not presented) of the contaminated material with a pristine control soil (w/w), and the final treatment was a control soil. Treatments were conducted in triplicate with ten worms being exposed per jar, for a total of thirty earthworms per treatment. Earthworms were placed on the surface of each of the test soils and no physical effort was made to force the worms to burrow into the soil. To monitor survival, the worms were emptied out of the jars and counted daily for the initial 14 days (acute toxicity test period) and weekly thereafter (subchronic toxicity testing period) for the following eight weeks. Soil was returned to the corresponding jar and the live worms were placed back on the surface. The assays were conducted under lighted conditions, to facilitate burrowing into the soil, and at a constant temperature (22 °C).

RESULTS AND DISCUSSION

Soils. Results from general soil characterization are provided in Table 1. Results indicate that soil texture ranged from loam to clay loam. Based on the results of particle size distribution, all soils (A, B, C and D) are classified as fine textured (greater than 50% by mass < 75 um). Soil textures according to the Canadian Society of Soil Science (1978) are Clay Loam for Soils A, B and C and Loam for Soil D and the Control Soil.

Physical and Chemical Characterization. Results from hydrocarbon characterization within biotreated soils are provided in Table 2. For comparison purposes, CCME PHC standards for fine textured surface and subsurface soils have also been provided (Table 2). PHC concentrations within all soils are in excess of CCME PHC standards for various soil types, land uses, and exposure pathways. Looking at specific fractions however, we see the PHC concentrations for F2 and F4 are below the CCME standards. However, the PHC F3 fraction remains elevated above CCME PHC surface soil standards in all four soils; assuming the relevant exposure pathway is ecosoil contact. Thus, it is the F3 fraction that causes the soils to fail for PHC under the current CCME standards.

PAHs were detected within Soils A, C and D but were below detection (< 0.05 mg/kg) within Soil B. Quantifiable PAHs ranged from 2-ring (naphthalene) to 4-ring compounds (chrysene); however, concentrations in Soils A, C and D were not appreciable.

Toxicity Testing. Results from the battery of toxicity tests are presented in Table 3. Results from the Microtox™ assay indicate that all soil extracts were non-toxic (both EC_{20} and EC_{50}) throughout the test period.

Seed germination assays indicate that ≥ 90 % of both radish and oats had germinated in all soils within four days. A seed germination rate of ≥ 90 % was used to indicate that the soils were non-toxic. Data for 50% diluted soil were not presented as 100 % contaminated soil initiated non-toxic responses in test organisms, as measured, and the 50 % dilution data did not significantly differ from the 100 % treatment.

Results from the earthworm survival bioassay (Table 3) indicate that earthworm survival was 100 % during both the acute (14 day) and chronic (10 week) phases of the experiment. Toxicity data for 50 % diluted soil (not presented) had a 100 % survival rate. There was no visible evidence of any behavioral impairment to the earthworms.

Discussion. The development of the CCME PHC standards involved the addition of fresh PHC to soil and the measured toxic response(s) to a variety of organisms at various levels of biological organization. The CCME Tier I PHC standards development protocol produced PHC criteria that are protective of ecological health impacts produced by recent spills or discharges to soil as this PHC is readily bioavailable and capable of eliciting a toxic response. The CCME Tier I PHC standards do not incorporate the transformation of hydrocarbons with weathering and the development of bioavailability and mass transfer limitations over time. As such the CCME Tier I standards may be overly-protective of the ecological risk actually posed by residual hydrocarbons in soil following biotreatment, at least for the sites investigated in this study using microbiological, seed germination and earthworm toxicity trials.

CONCLUSIONS

Based on the results of soil characterization, PHC was in excess of various CCME PHC standards indicating a perceived risk to environmental health. Based on the results of the three bioassays, however, there was no indication of toxicity, as measured by exposures to microbes, vegetation and earthworms. We conclude that the risk of toxicity from weathered PHC in bioremediated loam to clay loam soils may not be captured by current CCME PHC standards. In order to determine the actual risk of weathered PHC to various ecological receptors, toxicity tests should be performed on weathered PHC in conjunction with site-specific chemical characterization. The coupling of site-specific toxicity data with chemical characterization should assist in achieving remediation goals and standards that protect against actual risk.

ACKNOWLEDGMENTS

The authors thank Norwest Labs for providing chemical and physical analyses of soils used in this study. Thanks to Sumithrai Vasanthan who performed seed germination and earthworm bioassays. Komex International Ltd provided funding for this project.

REFERENCES

Alexander, M. 1995. "How Toxic are Toxic Chemicals in Soil?" *Environmental Science and Technology.* 29(11): 2713-2717.

Alexander, M. 2000. "Aging, Bioavailability, and Overestimation of Risk from Environmental Pollutants." *Environmental Science and Technology.* 34(20): 4259-4266.

ASTM. 1994. *Standard Practice for Conducting Early Seedling Growth Tests.* E1598–94. American Society for Testing and Materials. West Conshohocken, PA.

ASTM. 1995. *Standard guide for conducting a laboratory soil toxicity test with lumbricid earthworm Eisenia foetida.* E1676–95. American Society for Testing and Materials. West Conshohocken, PA.

Angehrn, D., R. Galli, and J. Zeyer. 1998. "Physicochemical Characterization of Residual Mineral Oil Contaminants in Bioremediated Soil." *Environmental Toxicology and Chemistry.* 17(11): 2168-2175.

Canadian Council of Ministers of the Environment (CCME). 2000. *Canada-Wide Standards for Petroleum Hydrocarbons (PHC) in Soil.* Canadian Council of Ministers of the Environment, Winnipeg, MB.

Canadian Soil Survey Committee, Subcommittee on Soil Classification. 1978. *The Canadian System of Soil Classification.* Canadian Department of Agriculture. Publ. 1646. Supply and Services Canada. Ottawa, ON.

Environment Canada. 1992. *Biological test method: Toxicity test using luminescent bacteria (Photobacterium phosphoreum)*. Environment Canada, Ottawa, Ontario. Report EPS 1/RM/24.

Greene, J.C., C.L. Bartels, W.J. Warren-Hicks, B.R. Parkhurst, G.L., Linder, S.A., Peterson, and W.E. Miller. 1989. *Protocols for short term toxicity screening of hazardous waste sites*. EPA 600/3 - 88 – 029, United States Environmental Protection Agency, Environmental Research Laboratory, Corvallis, OR.

Table 1. Physical and Chemical Properties of Pristine, Control and Biotreated Soils used in Study

	Soil A	Soil B	Soil C	Soil D	Control
Texture[a]	Fine	Fine	Fine	Fine	Fine
Sand	40.7 ± 1.3	34.7 ± 1.3	23.7 ± 0.7	47.1 ± 0.7	44.0
Silt	32.8 ± 1.7	38.3 ± 0.9	41.2 ± 0.5	30.2 ± 0.4	37.6
Clay	26.5 ± 0.9	27.0 ± 0.2	35.1 ± 0.1	22.7 ± 0.3	18.4
Organic Matter (%)	4.0 ± 0.1	5.8 ± 0.3	3.8 ± 0.1	2.1 ± 0.2	2.7
pH (Sat. Paste)	7.5 ± 0.0	7.5 ± 0.1	7.2 ± 0.0	7.2 ± 0.1	7.1
EC (dS/m)	1.47 ± 0.03	2.19 ± 0.08	0.96 ± 0.04	2.72 ± 0.09	0.53
SAR[b]	1.70 ± 0.0	3.00 ± 0.11	0.80 ± 0.00	1.10 ± 0.00	0.10
Bulk Density (g/cc)	1.20 ± 0.01	1.14 ± 0.01	1.18 ± 0.02	1.26 ± 0.01	1.26

[a] Fine textured soils are defined as having greater 50% by mass particles less than 75 um (D_{50} < 75 um).
[b] Sodium Adsorption Ratio.
Note physical and chemical properties are reported as the average of triplicate analyses ± 95 % confidence interval; the control values are the result of a single analysis.

Table 2. Petroleum Hydrocarbons (PHC) and Polycyclic Aromatic Hydrocarbons (PAHs) within Biotreated Soils Compared to CCME PHC Standards for Various Land Uses Assuming EcoSoil Contact as Exposure Pathway

	Agriculture	Residential/ Parkland	Commercial	Industrial	Soil A (mg/kg)	Soil B (mg/kg)	Soil C (mg/kg)	Soil D (mg/kg)
Petroleum Hydrocarbons								
Fraction 2 ($>C_{10}-C_{16}$)	900[a]/2200[b]	900[a]/2200[b]	1500[a]/3000[b]	1500[a]/3000[b]	289 ± 9.9	117 ± 28.0	153 ± 12.8	279 ± 39.1
Fraction 3 ($>C_{16}-C_{34}$)	800[a]/3500	800[a]/3500[b]	2500[a]/5000[b]	2500[a]/5000[b]	3693 ± 32.7	1127 ± 6.5	1113 ± 18.2	1993 ± 45.7
Fraction 4 ($>C_{34}-C_{50}$)	5600[a]/10000[b]	5600[a]/10000[b]	6600[a]/10000[b]	6600[a]/1000[b]	2180 ± 49.3	726 ± 4.3	921 ± 27.2	894 ± 44.2
PAHs[c]								
Naphthalene	0.1	0.6	22	22	0.09 ± 0.01	< 0.05	0.07 ± 0.00	0.10 ± 0.01
Phenanthrene	0.1	5	50	50	0.22 ± 0.01	< 0.05	0.05 ± 0.01	0.13 ± 0.01
Chrysene	-	-	-	-	0.12 ± 0.01	< 0.05	< 0.05	0.09 ± 0.00
Fluorene	-	-	-	-	0.10 ± 0.00	< 0.05	< 0.05	0.05 ± 0.00

All results reported on a dry weight basis.
CCME PHC F1 not determined as results from biotreatment indicated that C_5-C_{10} hydrocarbons were below detection.
[a] CCME PHC Standards for surface soil (< 1.5 m below ground surface).
[b] CCME PHC Standards for subsurface soil (> 1.5 m below ground surface).
[c] Only those PAHs that were detected are provided in Table.
Note: Detectable PHCs and PAHs are reported as the average of triplicate analyses ± 95 % confidence interval.

Table 3. Results of Microtox™, Seed Germination and Acute and Subchronic Earthworm Bioassays Following Exposure to Biotreated PHC Contaminated Soils

	Soil A	Soil B	Soil C	Soil D	Control
Earthworm (% Survival)[a]					
14 day	100	100	100	100	100
10 week	100	100	100	100	100
Plant (% Germinated)[*]					
Radish (day 4)	100	100	100	100	100
Oats (day 4)	96.7	93.3	90.0	96.7	100
Radish (day 7)	100	100	100	100	100
Oats (day 7)	100	100	93.3	100	100
Radish (day 11)	100	100	100	100	100
Oats (day 11)	100	100	93.3	100	100
Microtox (% Sample)[*]					
EC_{50} (5 minute)	>100	>100	>100	>100	ND
EC_{20} (5 minute)	>100	>100	>100	>100	ND
EC_{50} (15 minute)	>100	>100	>100	>100	ND
EC_{20} (15 minute)	>100	>100	>100	>100	ND

[a] 50% diluted soil data not shown, data are the same as in the 100 % contaminated soil.
ND = not determined.

A METHOD FOR ASSESSING THE FULL COSTS AND BENEFITS OF GROUNDWATER REMEDIATION

Paul E. Hardisty[1]
Jane Dottridge[2]
Steve Wallace[3]
Jonathan Smith[4]
Ece Ozdemiroglu[5]
Jonathan Fisher[6]

1. Komex Europe Ltd., Nicosia, Cyprus
2. Komex Europe Ltd., London. UK
3. Lattice Property Holdings Plc., Basingstoke, UK
4. Environment Agency, National Groundwater & Contaminated Land Centre, Solihull, UK
5. EFTEC Ltd., London. UK
6. Environment Agency, National Centre for Risk Assessment & Options Appraisal, London, UK

ABSTRACT: Economic analysis can be used to compare the real overall costs of pollution, including private costs (to the problem holder) and external costs (to society), and the benefits accruing from remediation and the avoidance of future damages. Overall decisions on the required level of remediation can thus be made with an understanding of the full economic ramifications, including the wider implications for the environment and society as a whole. Recognizing the need to include economic concerns in groundwater remediation decision making, the UK Environment Agency have recently published a research document which provides a method for assessing groundwater remediation alternatives. This framework is linked directly to existing UK guidance on risk assessment for contaminated sites. The method involves an iterative screening approach, starting with high level analysis of the costs and benefits of remedial objective options and the technical approaches which are best suited for reaching those objectives. The benefits analysis includes both private benefits, and if possible wider external benefits (often described as the value of damage averted by taking action). Valuation of external benefits of remediation is not straightforward, and can include the value of damaged resources, option values, and intrinsic worth (existence and bequest values). Often, only partial monetization of some benefits can be practically achieved. However, benefits can be compared with expected annualized costs for achieving overall specific remedial objectives, allowing a preliminary identification of the economically optimal objective. The framework provides a tool with which firms and regulators can negotiate a position that balances their respective concerns (social optimum against private optimum). Once the remedial objective has been set, and with it the most economically attractive approach, technology selection becomes merely a least-cost analysis. In many cases, the hidden costs of certain remedial objectives and approaches are revealed. This framework is applied to the problem of a contaminated manufactured gas plant (MGP) in the United Kingdom. The site was used for coal gas production from the turn of the last century until the late 1950's. Deep migration of coal tar NAPL into the fractured bedrock aquifer underlying the site has resulted in a significant dissolved phase plume. Risk analysis was used to identify and prioritize pollutant linkages, and the impacts on receptors were then quantified and monetized. A variety of remedial objectives are evaluated, and cost-benefit analyses conducted, using notional least-cost solutions, and partially monetized private and social benefits. The results show the variability in benefit-cost ratio which can occur when comparing widely differing remedial objectives, and the economic implications associated with the use of passive techniques such as natural attenuation.

INTRODUCTION

Despite the vast sums spent globally on managing groundwater pollution in recent years, relatively little research has been conducted into applying cost-benefit techniques to problems of groundwater remediation and protection (Hardisty and Ozdemiroglu,1999). The available literature has been produced by economists or by technical (scientific and engineering) experts, but shares little common ground. Not unexpectedly the economic literature deals mainly with valuation of groundwater and the external economic benefits of groundwater protection. Some work deals directly with groundwater remediation. The technical-scientific literature focuses on the application of specific techniques and technologies to groundwater problems, and deals almost entirely with remedial costs, cost-comparisons, and cost-effectiveness. The wider benefits of remediation are rarely discussed. Much of this work is of primary interest to problem holders, but even so, very little is available which discusses the private benefits of remediation (Hardisty and Ozdemiroglu, 1999). When considering highly mobile contaminants, the need for a complete analysis of the economics of groundwater contamination, and a rational, objective analysis of the full costs and benefits of remediation, becomes apparent.

The UK Environment Agency's duty to consider the financial aspects of remediation is a statutory requirement in England and Wales by virtue Section 39 of the Environment Act 1995. The act requires that the Agency should "take account of the likely costs and benefits" in deciding whether, and how, to exercise its statutory powers.

BACKGROUND AND TERMINOLOGY

Remedial objectives, approaches, and technologies. At the outset, it is important to distinguish between the different levels at which decisions need to be made about remedial actions. The current literature makes reference to "remedial approaches", "remedial options", and "remedial technologies", sometimes interchangeably, and often without clear definition. For groundwater, the distinction between objectives, approaches and technologies is vitally important. These are formally defined below:
- *Remedial objective* is the overall intent of the remediation project. Objectives could include the degree to which groundwater is to be remediated, the protection of specific receptors, or the elimination or reduction of certain unacceptable risks.
- *Remedial approach* is the conceptual manner in which the objective is to be reached, and is defined specifically in terms of the source-pathway-receptor (SPR) linkage component they address: source removal, pathway interruption, or source protection/isolation/modification.

- *Remedial technologies* are the specific tools which form the components of the approach. For example, physical containment (a pathway interruption approach) can be achieved through use of slurry walls, sheet pile walls, or liners, often in conjunction with groundwater pumping and treatment. Source removal can be achieved through excavation and on-site treatment of contaminated soils (by a variety of techniques), or through many available in-situ techniques. A remedial solution will often incorporate several different remedial technologies.

Remedial objectives should be known before detailed design (technology selection) occurs. The choice of a remedial approach is a critical intermediate step, which can be used both as a tool to help set objectives (by considering and comparing various approaches at the conceptual level), and to guide the selection of the technological components which will make up the final design.

LEVELS OF DECISION MAKING

The subject can be considered at three main levels: 1) *policy level* (government), which for the purposes of this discussion is considered to be set, and is not considered in detail; 2) *remedial objective level*, and 3) *technology selection level*. An intermediate step between the objective (2) and the most applicable technology (3), is the remedial approach. The remedial approach is the link between the objective and technology levels, and needs to be considered at both of these levels.

Policy Objective Level. Policy objectives are set by the government, and are not the subject of this paper. However, decisions on what to remediate, what to protect and what to sacrifice, must be generally guided by the policy of the day. Policy could include maximization of human welfare, for instance. In the UK and European Union, new Act (or Directives in the case of the EU) are subject to a high-level cost compliance assessment. This ensures that the requirements of the new law are considered affordable to the national economy as a whole.

Remedial Objective Level. Setting the remedial objective (or risk management objective) for a given groundwater contamination problem, should explicitly incorporate the standard risk assessment. Only a limited number of remedial (or risk management) objectives are available: the receptor is protected, impacts to the receptor are reduced or eliminated, the contamination is removed, or contamination is reduced to pre-determined regulatory levels, based on generic risk assessment. The remedial objective is the level at which the benefits of remediation are most readily and fundamentally determined. If a valuable receptor is protected, a benefit to society accrues. If a receptor is not protected, a damage results. In this framework, benefits are tied clearly to the fundamental objective, and the basic approach used to achieve it. The choice of whether to achieve a given objective using pump-and-treat, a bio-barrier, or monitored natural attenuation (for instance), has a direct impact on costs (including any dis-benefits associated with the method (such as release of off-gases to the atmosphere, for instance), but benefits remain essentially constant.

Remedial Approach. As discussed above, there are now literally dozens of groundwater remediation technologies available to achieve any specific technical outcome. This is the most detailed level of assessment of remedial costs. Often several different technologies, each designed to achieve a specific technical outcome, will be required to accomplish the remedial objective. Because so many technologies exist, which achieve such different technical outcomes, at such widely varying costs, explicit cost-benefit analysis could literally require comparison of dozens of cost (technology) options, many of which are designed to achieve very different technical outcomes. The link between the remedial objective, and the many pieces of technology available on the market today, is the remedial approach.

The remedial approach (sometimes called remedial strategy) does not focus on technology, but on ways of breaking the SPR linkage which causes (or is likely to cause) pollution or harm (damage). The list of possible remedial approaches is relatively short: remove or modify the source, cut the pathway, or protect/move/manage the receptor. Consideration of remedial approach can be very useful in streamlining the CBA process, since a limited number of approaches need be considered. In this way, it provides a link between remedial objectives, and the many technologies available. Also, the degree to which the linkage is broken, the timing of the action, and the spatial location at which the action is taken, are all variables that must be considered when choosing the approach, since they will affect both the costs and the benefits accrued. A constraints analysis can be undertaken to help assess which approaches are practicable. Preliminary, high level, costs can be assigned to each approach that can feasibly achieve the desired objective, and compared to the benefits of achieving the objective. This provides a relatively quick strategic analysis of costs and benefits of groundwater remediation, before proceeding to detailed technology evaluation and cost analysis.

Remedial Technology Selection Level . Remedial technology selection involves identification of the most cost-effective way of achieving a remedial objective. This requires detailed comparison of capital and operation and maintenance (O&M) costs for technologies, over a projected project life span. Costs associated with production of secondary damages would be incorporated into the cost analysis, such as air pollution arising from remedial process emissions. Application of various constraints provides a life-cycle cost analysis (Hardisty et al, 1998). By using the intermediary remedial approach step, an iterative analysis of the project is possible.

FRAMEWORK METHOD
A framework for considering the complete costs and benefits of groundwater remediation has been developed (Hardisty and Ozdemiroglu, 2000), based on a detailed economic model presented in Hardisty et al (1999). The method is based on a tiered, iterative system, whereby remedial objectives are determined first with a preliminary high-level analysis. The costs of achieving various objectives are determined by considering remedial approach options best capable of meeting the objectives. These cost are compared to the benefits of meeting the objectives. Once the economically optimal objective-approach combination is selected, the least cost way of implementing the approach is selected. The methodology is illustrated in the following example.

EXAMPLE – MANUFACTURED GAS (MGP) PLANT SITE IN THE UK

Background and Setting. A hybrid site has been developed for this study based on field data gathered from a number of industrial sites in the East Midlands of the United Kingdom. Studying a hybrid site has enabled real data and a variety of existing and potential groundwater remediation issues to be investigated.

Site Description. The hybrid site is a former MGP, gasworks, located close to a town center in the UK, and covers an area of approximately 2 hectares (ha). The site is situated in a shallow valley associated with the partially culverted River S which runs adjacent to the western site boundary. Topography slopes gently downwards towards the north-east, following the river's flow direction. Industrial development of the site commenced in the early 1900's and processing facilities changed location, design and size until the 1980's when industrial activities ceased. Since that time, a number of site investigations have been conducted to assess the contaminant situation and some soil remediation has been completed. A mixture of industrial, commercial and residential properties surround the site.

The generalized geological sequence is about 5 m of sand, gravel and made-ground, overlying 25 m of fractured Triassic sandstone aquifer, overlying a competent marl. The sandstone is a major aquifer in the region, used for water supply. Groundwater exists within the shallow sand and gravels at a depth of about 3mbgl, and in the sandstone – the two units are in hydraulic continuity.

Contaminant Distribution and History. Gas manufacturing operations over the years have left a legacy of significant subsurface contamination. Coal tar NAPLs are present over a significant part of the site, predominantly in areas used for coal gas manufacture (retort houses, tar tanks, purifiers). Tars are composed primarily of polynuclear hydrocarbons (PAHs), but also contain significant amounts of benzene and aliphatic hydrocarbons. Coal tar NAPL exists predominantly in the made ground and the sand and gravels. Coal tar DNAPL has also penetrated into the fractured sandstone bedrock to a depth of 25 m, based on coring and geophysical data, but has not penetrated the underlying Middle Permian Marl. Coal tar in the sandstone appears to be restricted to major vertical and sub-vertical fractures and to horizons of higher matrix permeability within the sandstone. NAPL contamination has resulted in migration of dissolved phase contaminants in groundwater towards the river and off-site in the fractured sandstone. Currently a 500 m long benzene plume exists in the sandstone aquifer. The plume appears to be stable at present. Modeling has shown that the presence of this plume eliminates 4600 m^3/day of potential groundwater abstraction from the aquifer, including the volume of aquifer currently contaminated, and the volume which would become contaminated by nearby wells capturing the plume over time.

Risk Assessment. A risk assessment carried out for the site identified eleven SPR linkages which exist at the site. These are listed in Table 1, below. The possible remedial objectives that could manage the risks associated with each pollutant linkage are also shown in Table 2. Objectives which are clearly inappropriate or which would fail to achieve legal requirements have been eliminated. Note the limited number of possible remedial objective categories (A through F) in Table 2.

Table 1 SPR Linkages and Remedial Objective Options

	RECEPTOR	PATHWAY	SOURCE	REMEDIAL OBJECTIVE OPTIONS
1A	River	NAPL via sediments	NAPL in sediments	C D E
1B	River	Aqueous phase via sediments	NAPL in sediments	B C D E
2	River	Aqueous phase via bedrock	NAPL in bedrock	B C
3A	Surface Water User	NAPL via sediments	NAPL in sediments	A D E
3B	Surface Water User	Aqueous phase via sediments	NAPL in sediments	A D E
4	Surface Water User	Aqueous phase via bedrock	NAPL in bedrock	A E
5	Aquifer and users	Aqueous phase via bedrock	NAPL in bedrock	A B E F
6A	Residents off-site	Vapor from NAPL in sediments	NAPL in sediments	A E
6B	Residents off-site	Vapor from NAPL in bedrock	NAPL in bedrock	A D E
7A	Workers on and off-site	Vapor from NAPL in sediments	NAPL in sediments	A D E
7B	Workers on and off-site	Vapor from NAPL in bedrock	NAPL in bedrock	A E

Table 2 Possible Remedial Objective Categories

A	Protect potential receptor to risk-based level
B	Reduce impact to current receptor to prevent unacceptable risk
C	Eliminate impact to current receptor
D	Remove all contamination (source)
E	Reduce contamination to risk-based level
F	Institutional control of water resource (prevent use and monitor)

Remedial Approach Analysis. Next, remedial approach options able to reach each of the remedial objectives short-listed in Table 1, above, were developed, considering site conditions and the results of the risk assessment. In this example, 20 different approaches (alone or in combination) were evaluated on a preliminary basis, considering the type of technologies which would be entailed and their overall cost and practicability at the site. By iterating back up to the objective level, a short-list was developed of feasible remedial approaches able to achieve the various objectives for the 11 SPR linkages. The description of the most appropriate approaches is provided in Table 3, and the most appropriate approach for each objective for each SPR-linkage is shown in Table 4. Note that in all, only eight remedial approaches (alone or in combination) have been found worthy of more detailed analysis by CBA for this site.

Table 3 Remedial Approach Alternatives

DESIGNATION	APPROACH DESRIPTION	RELATIVE COST
Source methods		
S1	Remove NAPL in sediments	High
S1P	Partial removal of NAPL in sediments	Moderate
S2P	Partial removal of NAPL in bedrock	Moderate
Pathway methods		
P1D	Contain dissolved phase in sediments	High
P2D	Contain dissolved phase in bedrock	High
P4N	Contain vapor at limit of NAPL plume	Moderate
Receptor Methods		
R1	Collect and treat discharge to river at river	Very high
Other		
N3MNA	monitor natural attenuation processes acting on plume	Low to moderate

Risk Assessment, Toxicity, and Treatment Costs

Table 4 Remedial Approach Short-List Matrix
{Most appropriate approach for each objective option is indicated by an X}

SPR LINKAGE	REMEDIAL OBJECTIVE OPTIONS	S1	S1P	S2P	P1D	P2D	P4N	R1	N3MNA
1A	C				X				
	D	X							
	E		X						
1B	B		X						
	C			X					
	D	X							
	E		X						
2	B					X			
	C							X	
3A	A		X						
	D	X							
	E		X						
3B	A		X						
	D	X							
	E			X					
4	A			X					
	E			X					
5	A			X					
	B			X					
	E			X					
	F								X
6A	A		X						
	E								X
6B	A		X						
	D					X			
	E								X
7A	A		X						
	D					X			
	E								X
7B	A			X					
	E								X

Benefits Analysis. The risk analysis for the site indicates that there are only four main benefit categories, if the damage avoided concept is followed. In this case, benefits of remediation accrue if damage to the river and its users is avoided, if damage to the aquifer and its users are avoided, and if damage to the property itself and the neighboring properties is avoided. The methodology does not incorporate directly the benefits of avoiding damage to human health, and so these are not directly monetized. There is regulatory control over the use of water for potable supply in the UK. If groundwater is polluted the source will be condemned, and consequently the damage is equivalent to the cost of replacing that source. There is no option to continue using the water supply in a way that would cause harm to human health. Instead, benefits of remediation to workers and off-site residents are valued by the increase in property value realized by remediation, or the elimination of negative effects on property values which accrue due to remediation of the site. The benefits of remediation that could be readily monetized for this study are listed in Table 5, below. There are other benefits that could be also discussed, and included in the analysis, such as positive public relations benefits to the problem holder in undertaking the remediation, and the avoidance of fines and prosecution. These and other benefits which are more difficult to monetize are left out of this analysis, but could be included in a more rigorous treatment of the example.

Table 5 Remedial Benefits

BENEFIT CATEGORY	ONE TIME BENEFIT (£)	ANNUAL DAMAGE AVOIDED (£/PA)	PLANNING HORIZON (yrs)	VALUATION METHOD
Prevention of river water quality degrading to the point where the next lowest UK river category is reached	N/a	Negligible < £ 0.01 M	20 yrs	Recreation value is minimal due to poor current river water quality, and location in industrial area. Water not used for abstraction or supply. No ecological significance currently.
Lost potential water production from portion of aquifer rendered unusable by contamination in bedrock	N/a	Approx £ 0.05 M/pa for first 10 years, and then £ 0.2 M/pa for next 10 years	20 years for full restoration of plume through natural attenuation once source is removed	Modeled lost potential water production due to presence of plume, multiplied by current commercial market value of water in UK
Sale of property as commercial site, once made suitable for such use	£ 1.05 M	N/a	One time benefit	Current market value of commercial property in this part of the UK
Recovery of property value in sites within the vicinity of the site, as a result of remediation	£ 0.40 M	N/a	One time benefit	Notional improvement of 10% in average property value in adjacent sites, based on current market values, due to elimination of "blight".

The costs of the various remedial approaches are provided in Table 6. These are based on current market costs in the UK. The presumed least cost technology, based on the author's experience in the market on many similar sites, was used to cost the approach. For a more complete analysis, a formal least-cost analysis of technology options can be performed for each approach, and the results used to check and if necessary modify the initial high level analysis.

Cost Benefit Analysis. Using the calculated partial benefits and the remedial costs for various approaches, an economic comparison can be made of the remedial objective options available for dealing with each of the 11 SPR linkages. Varying the start time of implementation also has an effect on the calculated benefits. Table 7 shows the analysis of costs and benefits assuming immediate implementation of remediation, using a 5% discount rate. For each approach and combination of approaches, the benefit-cost ratio (BCR) is calculated. BCR values above unity indicate net positive economic benefits. A 20 year planning horizon has been used.

Table 6 Remedial Approach Cost Analysis

REMEDIAL APPROACH	REMEDIAL TECHNOLOGY FOR COSTING	CAPITAL COST (£M)	ANNUAL OPERATION COSTS (M£)	OPERATION TIME (yrs)	ASSUMPTIONS
S1	Complete excavation and treatment of contaminated sediments	1.1	-	-	Soil washing on-site is used
S1P	Partial excavation of contaminated sediments	0.5	-	-	Material is land-filled
S2P	Partial removal of NAPL in bedrock using angled wells and pumping, surfactant flushing	0.5	0.1	10	Sufficient NAPL is removed to have positive impact on dissolved mass flux.
P1D	Contain dissolved phase and NAPL in sediments using slurry wall and pumping with treatment.	0.3	0.1	20	System must remain operational over long term, shallow source remains in place.
P2D	Contain dissolved phase in bedrock by installing hydraulic containment system	0.3	0.1	20	System must remain operational over long term.
P4N	Contain vapor at limit of LNAPL plume by installing soil vapor extraction system around site perimeter	0.25	0.05	20	20 yrs without NAPL removal, 2 yrs with.
R1	Collect and treat discharge to river at river, using cofferdam at river's edge	0.7	0.1	20	System must remain operational over long term, shallow source remains in place
N3MNA	Monitored natural attenuation of groundwater contamination over time, assuming MNA effective	0.1	0.05	20	Increased monitoring capability

ANALYSIS AND DISCUSSION

Examination of the CBA results in Table 7, and the remedial approach short-list matrix in Table 4, above, yields some interesting observations. First is that the benefit cost ratio is maximized when approach S1P (partial excavation of contaminated sediments) is used (BCR= 3.14). This represents the lowest ratio of cost to the problem holder to benefits to society (including the problem holder, who in this case achieves a significant benefit from selling a property which has become fit for commercial use). Using this approach would manage 6 of the 11 SPR linkages outright (Table 4). However, combining this limited source removal (S1P) with a programme of monitored natural attenuation (N3MNA) still provides a positive BCR (1.33), while perhaps more completely serving the interests of society as a whole. This choice of objective would reflect a decision to consider the distribution of benefits as a key parameter in decision making. Consulting Table 4, we see that this combined approach would address an additional three remedial objectives, and manage a total of nine risk linkages. Only SPR linkages 2 and 4, both involving the river would not be satisfied with this solution. However, the associated risk assessment points to the fact that relatively low mass flux to the river in the form of dissolved phase contamination, does not result in serious environmental impact. This is due to the poor current quality of the river, situated in a heavily industrialized area, and subject to other discharges along its reach. This situation could of course change over time, and it is the Agency's objective to improve surface water quality.

Table 7 Cost Benefit Analysis

REMEDIAL APPROACH	DESCRIPTION	COMPLEMENTARY APPROACH	BENEFITS WHEN APPLIED ALONE (NPV £M)	BENEFITS WHEN APPLIED TOGETHER (NPV £M)	TOTAL NPV COST ALONE/ COMPL (£M)	BCR ALONE COMPL
S1	Complete excavation and treatment of contaminated sediments	P2D Contain plume in bedrock; N3 MNA natural attenuation of remaining plume	1.57	1.94	1.10 3.37	1.43 0.58
S1P	Partial excavation of contaminated sediments	P2D Contain plume in bedrock; N3 MNA natural attenuation of remaining plume	1.57	1.74	0.5 2.77	3.14 0.63
S2P	Partial removal of NAPL in bedrock	S1P Partial removal of NAPL in sediments + N3MNA	0.17	1.94	1.27 2.25	0.14 0.86
P1D	Contain dissolved phase and NAPL in sediments		0.13	-	1.75	0.07
P2D	Contain dissolved phase in bedrock	S2P Partial removal of NAPL in bedrock	0.37	0.59	1.54 2.05	0.02 0.29
P4N	Contain vapor at limit of LNAPL plume by	S1P Partial removal of NAPL in sediments	0.2	1.77	0.64 1.14	0.31 1.55
R1	Collect and treat discharge to river at river,		0.13	-	1.95	0.07
N3MNA	Monitored natural attenuation	S1P Partial removal of NAPL in sediments	0.17	1.62	0.72 1.22	0.24 1.33

Other remedial approaches considered, such as an active attempt to remove some of the NAPL contained in fractured bedrock, are not economically attractive to society at this site. This approach entails high costs, for relatively few overall benefits (BCR=0.14). This analysis shows that in situations involving complex conditions, consideration of the wider costs and benefits of groundwater remediation can assist in setting realistic and appropriate remedial objectives.

ACKNOWLEDGEMENTS

This work was funded through the UK Environment Agency R&D programme, with additional financial support and access to data provided by Lattice Property Holdings Plc., and Komex. The views expressed here are those of the authors, and do not necessarily reflect the views or policies of either the Environment Agency or Lattice Property Holdings Plc.

REFERENCES

Hardisty, P.E. and Ozdemiroglu, E., 1999. *Costs and Benefits Associated with remediation of Contaminated Groundwater: A review of the Issues*. UK Environment Agency Technical Report P278.

Hardisty, P.E., and Ozdemiroglu, E. 2000. *Costs and Benefits Associated with remediation of Contaminated Groundwater: A Framework for Assessment*. Environment Agency Technical Report P279.

Hardisty, P.E., Bracken, R.A., and Knight, M. 1998. The Economics of Contaminated Site Remediation: Decision Making and Technology Selection. *Geol Soc. Eng. Geol. Sp. Pub.* 14, pp 63 - 71.

Hardisty, P.E., Kramer, E., Ozdemiroglu, E., and Brown, A. 1999. Economic analysis of remedial Objective alternatives for an MtBE Contaminated Aquifer. *Proc. Hydrocarbons and Organic Chemicals in Groundwater*, NGWA, Houston, Nov, 1999.

SELECTIVE MOLECULAR PROBES FOR DETECTING AROMATIC-COMPOUNDS DEGRADING BACTERIA IN ENVIRONMENTAL SAMPLES

M. Martin, M. Sanchez, C.Garbi, E. Ferrer (Dpt. Bioquímica y Biologia Molecular. Universidad Complutense, Madrid, Spain)
A. Gibello (Dpt. Patologia Animal I. Universidad Complutense)
J.L. Allende (Dpt. Fisica Aplicada. Universidad Complutense)
M.J. Martinez-Iñigo and C. Lobo (IMIA "El Encin", Apt. 127, Alcala de Henares, 28800 Madrid)
C. Martin (Dpt. Biología Vegetal, Universidad Politecnica, Madrid)

ABSTRACT: A set of ERIC primers was assayed for the detection of some aromatic compound-degrading bacteria in environmental samples. *Pseudomonas* GCH1 strain was used in contaminated soil bioremediation, and amplification by PCR using the DNA isolated from soil gave a specific DNA band pattern corresponding to the GCH1 strain. Based on 3,4-dihydroxyphenylacetate (3,4-DHPA) dioxygenase amino acid sequence and DNA sequence data of homologous genes, two different oligonucleotides were designed which have been assayed for detecting 3,4-DHPA related aromatic-compounds degrading bacteria in soil samples. The results indicated that these molecular techniques were highly specific and may represent a powerful tool for efficient remediation of xenobiotics pollutants by natural microbial communities.

INTRODUCTION

The efficient aplication of bioremediation projects remains to understand the behaviors of microorganisms involved in the degradation or removal of the target pollutants at the remediation sites. Aromatic compounds-degrading bacteria are most often used in these bioremediation systems, because aromatic compounds are particularly stable and prevalent in nature. They enter the environment through both natural processes and the industrial and agricultural pursuits of mankind. Catabolic pathways for aromatic compounds include three main steps (i) hydroxylation of the aromatic ring, (ii) ring cleavage, and (iii) formation of end-products which are central pathways intermediates, such as succinate, acetil-CoA or pyruvate. We have studied the degradative patways of different aromatic compounds, 4-hydroxy-phenylacetate (HPA), 4-hidroxybenzoate, propachlor and oxadiazone (Martín et al., 1991; 2000). The 4-HPA degradative operon of *Klebsiella pneumoniae* have been partially cloned and some of its enzymes characterized. 4-HPA hydroxylase (HPA-hyd) and 3,4-dihydroxyphenylacetate dioxygenase (DHPA-diox), are the two first enzymes involved in the ring cleavage (Gibello et a., 1994; 1997).

In the present study, we have tested PCR primers for the detection of bacteria used in bioremediation processes. Also, we have designed specific oligonucleotide probes by using Fluorescence In situ Hybridization (FISH) as potential tools for the identification of ring cleavage genes in degrading bacteria.

MATERIALS AND METHODS

Bacterial strains, plasmids and conditions of growth. *Escherichia coli* CC118 (pAG464) was obtained by transformation of *E. coli* CC118 with pAG464, a derivative of pUC18 plasmid containing the *K.pneumoniae hpa*B gene (Gibello et al. 1994). Cells were grown aerobically at 37°C in LB medium and Ampicillin (Ap) was added at the final concentration of 100 µg/ml,
Pseudomonas GCH1 (Martín et al., 2000) was used in the soil bioremediation assays.

Sequence analysis. The complete nucleotide sequence the *hpa* genes from *K. pneumoniae* on pAG464 was determined from both strands by the dideoxy-chain-termination method (Sanger et al. 1977) and submitted to the EMBL Nucleotide sequence data base with the accession number AJ000054. Computer analysis of the DNA and amino acid sequence data was carried out by using the GCG software package.

DNA isolation from soil bacteria. Trays contained 700g of soil were inoculated with LB liquid culture of GCH1 bacterial strain at the exponential phase of growth. Two different inoculum doses of 100mL and 150mL were applied. After 1-day incubation at 25 °C and 60% air humidity, DNA was extracted from 500-mg soil samples using FastDNA spin kit for soil (BIO 101, Carlsbad CA).

PCR and FISH Assays. PCR was performed in a total volume of 25 µL. The final reaction mixture contained 1 unit of Ecotaq DNA polymerase (Ecogen,S.R.L.), deoxynucleoside triphosphates at a concentration of 400µM, 1x Ecotaq buffer and 2mM Mg_2Cl solution. The repetitive intergenic consensus (ERIC) sequences were used to design PCR primers (Little et al., 1998). Before amplification, samples were heated at 94°C for 5 min. Then they were subjected to 36 cycles of denaturation at 94°C for 1 min, annealing at 52°C for 1 min and extension at 72°C for 2min, followed by a final extension at 72°C for 5 min. The 25µL reaction was separated by 2% agarose gel electrophoresis and the products visualised by UV light after ethidium bromide staining. Fish assays were carried out at 43°C, according to Duteau et al. (1998).

RESULTS AND DISCUSSION

DHPA diox sequence. Computer analysis of the *K. pneumoniae* DNA fragment cloned in pAG464, revealed four open reading frames (ORFs) with a high nucleotide sequence identity to the *E.coli hpa* genes encoding the 5-carboxymethyl-2-hydroxy-muconic semialdehyde dehydrogenase, 3,4-dyhydroxyphenylacetate 2,3-dioxygenase, 5-carboxymethyl-2-hydroxy-muconic acid isomerase and 2-oxo-hept-3-ene-1,7-dioic acid hydratese. The second ORF corresponding to the *hpa* B gene, that starts at position 798 bp, encoded the DHPA diox. which showed a high overall identity with its homologous enzymes of *E. coli*, strains C and W (Roper and Cooper 1990; Prieto et al. 1996), and *Salmonella dublin* (DHPA Sd: accesssion number AF144422). Since the comparison of these enzymes to the DHPA diox from *Pseudomonas aeruginosa* (DHPA Pa;Stover et al., 2000), also revealed a high grade of identity (63%), this type of enzyme seems to be highly conserved at least among these enterobacterial species and *Pseudomonas* sp (Fig. 1).

```
              1                                                    50
DHPAecC   ~~MGKLALAA KITHVPSMYL SELPGKNHGC RQGAIDGHKE ISKRCR....
DHPA Sd   ~~MGKLALAA KITHVPSMYL SELPGKNHGC RQAAIDGHIE IGKRCR....
DHPAecW   ~~MGKLALAA KITHVPSMYL SELPGKNHGC RQGAIDGHKE ISKRCR....
DHPA Kp   ~~MGKLALAA KTTHVPSMYL SELPGKNHGC RQGAIDGHKE IGKRCR....
DHPA Pa     MGKVALAA KITHVPSLYL SELPGPRHGC RQPAIDGHRE IGRRCR....
AMPD Pp   MATGEIISGF LAPHPPHMLY AENPPQNEPR SNGGWEQLRW AYERARASVE
AMBD Ps   ~MGKIVAAG  GTSHI...LM SPKGCEESAA R..VVNGIAE LGRRLK....
              51                                                  100
DHPAecC   EMGVDTIIVF DTHWLVNSAY HINCADHFEG VYTSNELPHF IRDMTYNY.E
DHPA Sd   EMGVDTIIVF DTHWLVNSAY HINCADHFQG VYTSNELPHF IRDMTYDY.D
DHPAecW   EMGVDTIIVF DTHWLVNSAY HINCADHFEG VYTSNELPHF IRDMTYNY.E
DHPA Kp   ELGVDTFIVF DTHWLVNSAY HINCADHFQG VYTSNELPHF IRDMTYDY.D
DHPA Pa   ELGVDTIVVF DTHWLVNAGY HINCAPHFEG LYTSNELPHF IANMEYGF.P
AMPD Pp   ALKPDVLLVH SPHWITSVGH HFIGVPELSG RSVDPIFPNL FR.FDYSM.K
AMBD Ps   EARPDVLVII TSDHMFNI.. NLSMQPRFVV GIADSYTPMG DMDIPRDLVP
              101                                                 150
DHPAecC   GNPELGQLIA DEALKLGVRA KAHNIPSLKL EYGTLVPMRY MNEDKHFKVV
DHPA Sd   GNPELGHLIA DEAVKLGVRA KAHNIPSLKL EYGTLVPMRY MNSDKHFKVV
DHPAecW   GNPELGQLIA DEALKLGVRA KAHNIPSLKL EYGSVLVPMRY MNEDKRFKVV
DHPA Kp   GNPELGHLIA DKTVKLGVRA KAHNIPSLKL DYGTLVPMRY MNADKHFKVV
DHPA Pa   GNPELGRILA EGCNALGVET LAHDATTLG.  EYGTLVPMRY MNQDRHFKVV
AMPD Pp   IDVDLAEACY EEGRNVGLET KMMRNPRFRV DYGTITTLHM IRPQWDIPVV
AMBD Ps   GSREVGRAIA LQDADEDGF.. DLCQAEEYSL DHGIMIPILF MGM.KEIPVV
              151                                                 200
DHPAecC   SISAFCTVH. .....DFADS RKLGEAILKA IEQY..DG.T VAVLASGSLS
DHPA Sd   SISAFCTVH. .....DFADS RRLGEAILKA IEKY..DG.T VAVFASGSLS
DHPAecW   SISAFCTVH. .....DFADS RKLGERIVKA IEQY..DG.T VAVLASGSLS
DHPA Kp   SISAFCTVH. .....DFADS RKLGEAIRKA IEKY..DG.T VAVLASGSLS
DHPA Pa   SVSALCTVHY .....LNDSA RLGWAMRKAV EEHY..DG.T VAFLASGSLS
AMPD Pp   SISANNTPYY LSMEEGLTEM DLLGKATLEA VRKS...GKR AVLLASNTLS
AMBD Ps   PVIVNINTDP IPSARRCV.. .ALAESIRQA IEKRTPDGCR VAVVGAGGLS
              201                                                 250
DHPAecC   H.RFIDDQRA EEGMNSYTRE ..FDRQMDER VVKLWREGQF KEFCNMLPEY
DHPA Sd   H.RFIDDQRA EEGMNSYTRE ..FDHQMDER VVKLWREGKF KEFCTMLPEY
DHPAecW   H.RFIDDQRA EEGMNSYTRE ..FDRQMDER VVKLWREGQF KEFCNMLPEY
DHPA Kp   H.RFIEDQRA EEGMNSYTRE ..FDHQMDER VVKLWREGKF KEFCTMLPEN
DHPA Pa   H.RFAQNGQA PDFSDRIWSP ..FLEVLDHE VVQMWQEGRW AEFCGMLPEY
AMPD Pp   HWHFNQEPEP PEDMTKEHPE SLAGYQWDMR MIDLMRRGQM QEVFRLLPQF
AMBD Ps   HWLCVP.... ........RH GEVSEKFDHM VMDELARGNA EELVAMGNEA
              251                                                 300
DHPAecC   ADYCYGEGN. .....MHDTV .......... MLLGMLGWDK YDGKVEFITE
DHPA Sd   ADYCYGEGN. .....MHDTV .......... MLLGMLGWDK YDGKVEFITD
DHPAecW   ADYCYGEGN. .....MHDTV .......... MLLGMLGWDK YDGKVWSLSP
DHPA Kp   AEYCYGEGN. .....MHDTV .......... MLLGLLGWDK YDGKEWNLSP
DHPA Pa   ASKGHGEGF. .....MHDTA .......... MLLGALGWSA YDGKAEVVTP
AMPD Pp   IEESFAEVKS GAFTWMHAAM QYPE...... LAAELHGYGT VIGTNAVME
AMBD Ps   IIDQGGNAGV EILTWIMAAV ASEASSGEKV FYEAMTQWFT GIGGMEFHVK

              301
DHPAecC   .LFPSSGTGQ VNAVFPLPA ~~~~~~~~~~ ~~~~~~~~~~ ~~~.  87.3%
DHPA Sd   .LFASSGTGQ VNAVFPLPA ~~~~~~~~~~ ~~~~~~~~~~ ~~~.  89.4%
DHPAecW   .SYSQASWHR SG~~~~~~~ ~~~~~~~~~~ ~~~~~~~~~~ ~~~.  87.6%
DHPA Kp   NCLPASGTRP G~~~~~~~~ ~~~~~~~~~~ ~~~~~~~~~~ ~~~.
DHPA Pa   .YFGSSGTGQ INAVFPVTA QDGSAIPAAQ AGNPAGASC ASRL.     63%
AMPD Pp   WNLVKAGLGR SAAAAR~~~ ~~~~~~~~~~ ~~~~~~~~~~ ~~~.    26%
AMBD Ps   ~~~~~~~~~~ ~~~~~~~~~~ ~~~~~~~~~~ ~~~~~~~~~~ ~~~.  26%
```

Figure 1. Comparisson of DHPA-diox sequences from different sources. Underlined sequences are identical to DHPA sequence from *Thermoplasma acidophillum*.

In absence of sequencing data from other dioxygenases from other species (eg. *Flavobacterium* sp.), this comparison (Fig. 1) suggest a common evolutionary origin for the DHPA diox. in gram negative bacteria very different to that established for the homologous enzymes isolated from Gram positive bacteria belonging to the major extradiol dioxygenase family (Boldt et al., 1995).

In comparison to other aromatic dioxygenases so far sequenced, DHPA diox. also showed significant similarity (34.5%) to the 2-aminophenol-1,6-dioxygenase (β subunit) of *Pseudomonas pseudoalcaligenes* (AMPD Pp;Davis et al., 1999) and 27.6% identity in 221 aa overlap, to the aminobifenyl 2,3-diol 1,2-dioxygenase of *Pseudomonas stutzeri* (AMBD Ps;Ouchiyama et al. 1998).

So based on the comparison of amino acid sequence between the all DHPA diox with the other aromatic dioxygenases (Fig.1), we have designed two different oligonucleotide probes for using as probes in FISH assays:

Diox 1:   5´ - GTGGTNTCGNTCTCCGC - 3´

Diox 2:   5´ - TCNGCCTTCTGTACCGTT - 3´

GCH1 Strain Identification by PCR assays. Figure 2a showed the specific DNA amplification patterns for pure GCH1 culture in LB liquid medium. When DNA was extracted from soil samples inoculated with GCH1 strain, PCR pattern shared 3 fragments (800 pb, 1100 pb and at between 1500 and 2072 bp-size fragments) and exhibited an extra fragment of 500 pb (Figure 2c,d) which is most likely due to amplification products from soil native microorganism DNA.

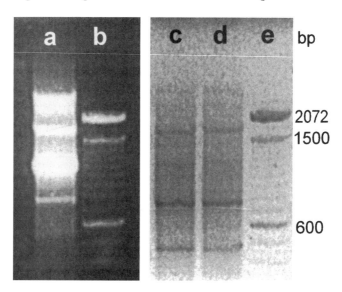

Figure 2: PCR results obtained by using ERIC sequences as primers for the amplification of the genomic DNA from GCH1 strain isolated from: (Lane a) pure bacterial culture in LB liquid medium and (Lanes c, d) soil samples inoculated with GCH1 at different doses. Lanes b, e: 100 bp DNA ladder marker (GibcoBRL) (2µL).

The results indicated that ERIC sequences permit the identification of pesticide degrading bacteria within the soil microbial population. Consequently, specific PCR are proposed for monitoring the bacteria introduced in contaminated soils submitted to remediation by bioaugmentation techniques.

CONCLUSIONS

PCR method by using the ERIC probes has showed to be an efective tool for the detection of specific bacterial strains used in bioremediation processes.

Specific oligonucleotides based on conserved sequences from genes involved in aromatic compounds catabolism, could be used for developing FISH protocols: genotypic catabolic potential could be investigated.

Molecular techniques have showed to be efective tools for the identification of bacteria in environmental samples.

ACKNOWLEDGMENTS

This work was partially funded by grant CAM07M/0030/1999 (Comunidad Autónoma de Madrid, Spain).

REFERENCES

Davis JK, He Z, Somerville CC and Spain JC. 1999. "Genetic and biochemical comparison of 2-aminophenol 1,6-dioxygenase of *Pseudomonas pseudoalcaligenes* JS45 to meta-cleavage dioxygenases: divergent evolution of 2-aminophenol meta-cleavage pathway". *Arch. Microbiol.* 172:330-339.

DuTeau N.M, Rogers J.D., Bartholomay C.,and K.F. Reardon. 1998. "Species-specific oligonucleotides for enumeration of *Pseudomonas putida* F1, *Burkholderia* sp. Strain JS150, and *Bacillus subtilis* ATCC 7003 in biodegradation experiments". *Appl. Environ.Microbiol.* 64: 4994-4999.

Gibello,A., Martin,M., E.Ferrer, and A. Garrido-Pertierra. 1994. "3,4-Dihydroxyphenylacetate 2,3-Dioxygenase from *Klebsiella pneumoniae*. A magnesium dioxygenase involved in aromatic catabolism". *Biochem. J.* 301:145-150.

Gibello, A, Suárez, M., Allende, J.L. and M. Martín. 1997. "Molecular cloning and analysis of the genes encoding the 4-hydroxyphenylacetate hydroxylase from *Klebsiella pneumoniae*". *Arch. Microbiol.* 167: 160-166.

Little E. L., Bostock R. M. and B. C. Kirkpatrick 1998. "Genetic characterization of *Pseudomonas syringae* pv. syringae strains from stone fruits in California". *Applied and Environmetal Microbiology*: 3818-3823.

Martín, M., Gibello, A., Fernández, J. , Ferrer, E. and A. Garrido-Pertierra.1991. "Catabolism of 3- and 4-hydroxyphenylacetic acid by *Klebsiella pneumoniae*". *J. Gen.Microbiol* 132: 621-628.

Martín, M., Mengs, G., Plaza, E., Garbi C., Sánchez M. Gutierrez F., Gibello, A. and E.Ferrer. 2000. "Propachlor removal by a soil isolated *Pseudomonas* strain in immobilized cell system. *Appl. Environ Microbiol.* 40: 34-39.

Prieto MA and Garcia JL. 1996. "Molecular characterization of the 4-Hydroxyphenylacetate catabolic pathway of *Escherichia coli* W: engineering a mobile aromatic degradative cluster". *J. Bacteriol.* 178:111-120.

Roper, DI and Cooper RA. 1990. "Subcloning and nucleotide sequence of the 3,4-dihydroxyphenylacetate 2,3-dioxygenase gene from *E. Coli*" *C. FEBS Lett.* 275:53-57.

Sanger, F., Nicklen S. and A.R.Coulson. 1977. "Nucleotide sequencing with chain-terminating inhibitors". *Proc. Natl. Acad Sci.* USA 74:5463-5467.

Stover CK et al. 2000. "Complete genome sequence of *Pseudomonas aeruginose* PA01, an opportunistic pathogen". *Nature* 406: 959-964.

MONITORING MICROBIAL COMMUNITY CHANGES IN OIL-CONTAMINATED LABORATORY SOIL MICROCOSMS

K. Vanbroekhoven and R. De Mot (KUL, Leuven, Belgium)
L. Bastiaens, *J. Gemoets* and D. Springael (Vito, Mol, Belgium)

ABSTRACT: *In situ* bioremediation of oil-contaminated soils by means of bioventing is seen as a cost effective technology for remediation of oil contaminated sites. Laboratory microcosms containing a long-term heavy oil contaminated soil were treated as a bioventing simulation system, applying different system parameters (i.e., the amendment of different nutrient compositions: mineral nutrients and a commercial oleophylic nutrient) with or without oxygen. In addition, in some microcosms extra diesel fuel was added in order to examine the influence of the oil concentration on the developing microbial populations. The changes in microbial populations were followed by means of Denaturing Gradient Gel Electrophoresis (DGGE) of 16S rDNA fragments generated by PCR using general eubacterial primers, and by CFU counts after microbial extraction. Analysis of the hydrocarbon content was performed to investigate the biodegradation capacity. Bacteria capable of using high oil concentrations as a C-source, were isolated and characterised for their hydrocarbon utilization pattern. Based on the clustered DGGE patterns, the CFU counts, and the hydrocarbon content, an effect on the microbial oil degrading soil population could be observed for those soil samples to which nutrients were added. Aeration did not seem to make any difference.

INTRODUCTION

Contamination of soils and groundwater by various petroleum hydrocarbons is widespread throughout Europe and the rest of the world. The use and storage of fuels like gasoline and diesel fuel, has resulted in surface spills and leakage from numerous underground storage tanks. Remediation of petroleum contaminated soils historically focused on either excavation and disposal or treatment of contaminated soils, or 'pump-and treat' techniques. Both approaches have been found to be extremely costly (Bowlen and Kosson, 1995). Since the last decade, in *situ* bioremediation techniques such as bioventing and biosparging are considered cost-efficient technologies for the remediation of oil contaminated soils. Nevertheless, information concerning the response of the endogenous bacterial populations on the applied system parameters is rare. One of the determinating factors for applying these technologies is the possible existence of an oil floating layer, which often has to be removed before bioremediation can be applied.

Objective. The objective of this study is to gain insight in the population dynamics of LNAPL degrading bacterial communities in LNAPL contaminated soils. PCR-DGGE and CFU counts were used to study the evolution of and changes in microbial communities in oil contaminated soil samples (8,000 mg/kg dm oil) treated in microcosms that simulated an *in situ* bioventing system. The response of the microbial population to different applied system parameters (i.e., provision of nutrients and aeration) was examined. Moreover specialized bacterial populations, able to resist and grow on high oil concentrations, will be enriched in view of future biobarrier applications.

Two nutrient formula were used: a mineral nutrient formula containing N & P as NH_4NO_3 and $K_4P_2O_7$ solutions providing a final C:N:P ratio of 100:10:1; and an oleophilic nutrient (Inipol EAP-22 accelerator; Elf Atochem) (The most well-known example of the use of Inipol EAP-22 is from the Exxon Valdez accident, where biodegradation of hydrocarbons spilled on the coastline was accelerated using Inipol EAP-22.) This nutrient was added to the soil samples at 10% by weight compared to the hydrocarbon content as recommended by the manufacturer Elf Atochem. The effect of high oil concentrations was examined by amending the soil with 20,000 mg/kg dm oil containing 112.5 mg/kg dm PAH's (25 ppm phenantrene, 25 ppm pyrene, 25 ppm fluorene, 25 ppm fluoranthene and 12.5 ppm anthracene) or by adding 70,000 mg/kg dm oil containing the same concentration of PAH's.

The changes in microbial populations were monitored using DGGE, based on the separation of 16S rDNA fragments amplified by PCR, using a general eubacterial primer set (Heuer and Smalla, 1997). CFU counts were performed to count heterotrophic bacteria and oil degraders. Mineral oil contents were determined by GC-FID and PAH contents by GC-MS.

MATERIALS AND METHODS
Preparation of soil samples and set up of bioventing microcosms. 15 kg soil, historically contaminated with approximately 8,000 mg/kg dm mineral oil and originating from the floating layer-groundwater table interphase, was homogenized during 30 minutes in a concrete mixer, which had been previously cleaned for 2 hours with glass fragments and non-contaminated soil to prevent contamination of the target soil with metals. A part of the soil was withdrawn from the mixer to provide Fraction 1, i.e., soil without oil amendment. Diesel fuel containing PAH's was added to the remaining part to reach a final mineral oil and PAH content in the soil of respectively 20,000 mg/kg dm and 112.5 ppm (25 ppm pyrene, 25 ppm phenantrene, 25 ppm fluorene, 25 ppm fluoranthene and 12.5 ppm antracene). After 30 minutes of additional mixing, soil Fraction 2 was removed from the mixer to provide the extra contaminated soil samples. An additional amount of diesel fuel was added to the remaining fraction so that a concentration was reached of 70,000 mg/kg dm diesel fuel and a PAH contamination of 112.5 ppm (no extra PAH addition) (Soil Fraction 3).

Microcosms consisted of glass bottles (250 ml) filled with 300 g of the different soil fractions. Bottles subjected to aeration, were connected to an aeration system. Microcosms were aerated during 15 minutes every 4 days at a rate of 17l/hour/bottle. In order to supplement the soil samples in the microcosms with mineral nutrients (M.N.) at a ratio C:N:P 100/10/1, 3 solutions were prepared corresponding to the 3 soil fractions with different oil concentrations consisting of

NH_4NO_3 and $K_4P_2O_7$ dissolved in sterile H_2O. To monitor and correct for any abiotic degradation of hydrocarbons in the soil microcosms, microcosms were included in which the microbial community was poisoned by adding 2 ml of a formaldehyde/NaN_3 solution at a final concentration of 0.175 % formaldehyde and 0.065 % NaN_3. An overview of applied conditions is given in TABLE 1. Each condition was started in duplicate. The soil samples were analysed at the start of the experiment, after 32 days, 82 days, and 140 days of treatment. For DGGE analysis, 2 soil samples (a,b) were analysed from each bottle.

TABLE 1: Conditions applied to the soil microcosms in order to study the dynamics of LNAPL associated bacterial communities.

Sample	Condition
0	soil+ NaN_3 & formamide
1	soil
2	soil+ 20,000 mg/kg dm oil
3	soil+ 70,000 mg/kg dm oil
4	soil+ Air
5	soil+ 20,000 mg/kg dm oil + Air
6	soil+ 70,000 mg/kg dm oil + Air
7	soil+ Mineral Nutrients
8	soil+ 20,000 mg/kg dm oil + Mineral Nutrients
9	soil+ 70,000 mg/kg dm oil + Mineral Nutrients
10	soil+ Mineral Nutrients + Air
11	soil+ 20,000 mg/kg dm oil + Mineral Nutrients + Air
12	soil+ 70,000 mg/kg dm oil + Mineral Nutrients + Air
13	soil+ Inipol EAP22
14	soil+ 20,000 mg/kg dm oil + Inipol EAP22
15	soil+ 70,000 mg/kg dm oil + Inipol EAP22
16	soil+ Inipol EAP22+ Air
17	soil+ 20,000 mg/kg dm oil + Inipol EAP22+ Air
18	soil+ 70,000 mg/kg dm oil + Inipol EAP22 + Air

DNA-extractions, followed by PCR using general eubacterial primers. Isolation of DNA from soil samples was carried out according to an adapted protocol of the bead-beat method described by El Fantroussi et al. (1997). Using the extracted DNA as template, 16S rRNA gene fragments were amplified by PCR using the general 16S rDNA eubacterial primer set GC-63F and 518R (Marchesi et al., 1998). All reactions were performed with a Thermocycler (Biometra) in a reaction volume of 100µl. The thermal profile of the PCR reaction consisted of an initial denaturation step at 94°C for 5 min., 40 cycles of 1 min. denaturation at 94°C, primer annealing for 1 min. at 55°C and extension for 1 min. at 65 °C, and a final extension step of 5 min. at 65 °C.

Denaturing Gradient Gel Electrophoresis (DGGE). The PCR amplified 16S rDNA fragments of equal length were separated using DGGE (D-code, Biorad). A 8% (w/v) polyacrylamide gel (acrylamide/ bisacrylamide 37.5:1) was used with a denaturing gradient ranging from 40% to 60%, resulting from the addition of formamide and ureum. 100% denaturant contained 7M urea and 40% formamide. The polymerisation of the acrylamide was realised by adding APS 10% and TEMED leading to a final concentration of 0.09 %. An 8% (w/v) AA stacking gel was used on top of the DGGE gel to avoid denaturation of DNA fragments in the wells. 10 µl DNA sample was mixed with the same amount of DGGE loading buffer (0.25 ml 2% bromophenol bleu, 0.25 ml 2% xylene cyanol, 7 ml 100% glycerol and 2.5 ml distilled H_2O) prior to application on the gel. The electrophoresis was run at 60°C for 5 hours at 160 V. After electrophoresis the gel was fixed in 200 ml 0.5% acetic acid during 2 to 3 minutes and stained during 30 minutes in 100 ml 1xTAE buffer with 10 µl 10,000x SYBR green I solution.

Bacterial extraction from soil samples and CFU counts. Sterile flasks were filled with 2 g soil and 18 ml bidistilled H_2O was added. The suspensions were shaken for 2 hours at 250 rpm at 12°C. After 15 minutes of sedimentation of soil particles, the supernatans were diluted 10-fold and plate countings were conducted. The diluted samples were plated on rich and mineral medium with diesel fuel as the only C-source. After several days of incubation at 28°C, CFU were counted.

Analysis of PAH and mineral oil content. Soil samples were dried and homogenized by adding an equivalent amount of celite. The PAH's and mineral oil present in the dried soil were extracted using the Accelerated Solvent Extractor (Dionex ASE 200). The extraction solvent used for PAH's was THF/C_6 (Tetrahydrofuran/hexane, Merck) 80/20 and the extraction was performed at a pressure of 140 bar and a temperature of 100°C. After clean-up of the PAH extract, the final PAH-nonane mixture was injected via autosampling (Combipal) on the GC-MS (GC 8000 top, Interscience; MS Voyager, Thermoquest). Preliminary experiments, performed to optimise the PAH recovery from contaminated soil samples, showed recoveries of 86% for spiked anthracene and 98% for added pyrene.

The ASE extraction performed to determine the mineral oil content was carried out using aceton/hexane (Merck) 50/50 as extraction solvent, at a pressure of 140 bar and a temperature of 100°C. The mineral oil extracts, collected in the extraction tubes, were purified and stored at 4°C until injection on GC. Just before the injection, a clean-up with florisil at a ratio 3.33/1 hexane extract/florisil was performed. Finally, the extracts were analysed by GC-FID (Iwaco, Rotterdam). In preliminary experiments, an optimal recovery percentage of 65% of the added diesel fuel was found.

RESULTS AND DISCUSSION

Cluster analysis of the obtained PCR-DGGE patterns revealed 4 almost identical major clusters after 32 days (T1) and 82 days (T2; Figure 1) of treatment. A clear influence of both the addition of diesel fuel and nutrient

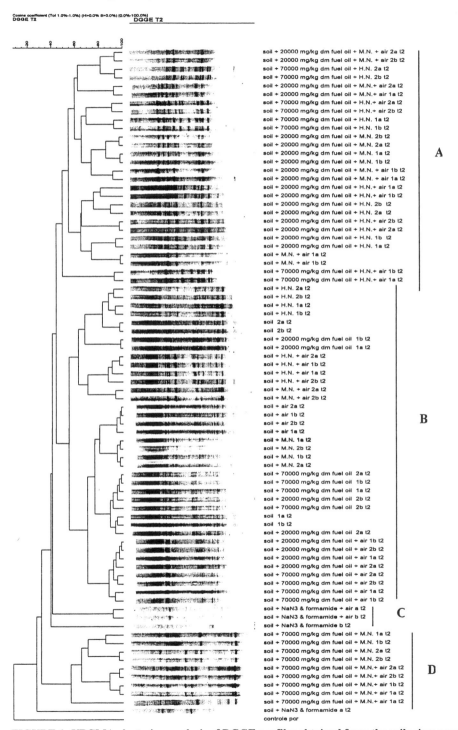

FIGURE 1: UPGMA clustering analysis of DGGE profiles obtained from the soil microcosms after 82 days of treatment (T2). Clustering using the cosine correlation coefficient was performed using Bionumerics (Applied Maths, Belgium) and expressed as percentage similarity. The different conditions are indicated. 1 & 2 represent the two duplicates for each treatment condition; a & b indicate DGGE patterns from 2 samples from the same microcosm.

amendment on the microbial community pattern, obtained by PCR-DGGE, could be deduced from the cluster analysis. The effect was most significant for soils provided with 70,000 mg/kg dm oil and mineral nutrients, where a shift in the DGGE patterns could be observed towards a more GC rich bacterial population. The observed clustering of community patterns obtained from soil samples with nutrient amendment (without oil addition) and samples with oil amendment (20,000 mg/kg dm and 70,000 mg/kg dm) with no further treatment together with the soil samples analysed at T0 indicated that little had happened in these soil microcosms after 32 and 82 days of treatment. Finally, both at T1 and T2, the impact of aeration on the microbial community patterns seemed negligible.

After 140 days (T3) of treatment, major changes took place regarding DGGE cluster analysis (Figure 3). A rearrangement of clusters occurred, although still 4 major clusters could be observed. The influence of both oil and nutrient amendment still remained visible, but the most pronounced effect, observed for the treatment with 70,000 mg/kg dm oil and mineral nutrients, had faded. DGGE patterns originating from both soil samples with addition of 20,000 mg/kg dm oil and mineral nutrient amendment (both with and without aeration) and soil samples with 70,000 mg /kg dm oil and hydrophobic nutrient addition now clustered together with the profiles from soils with 70,000 mg/kg dm diesel fuel and mineral nutrients.

On the contrary, a clear separate cluster included now the soils obtained from microcosms amended with 70,000 mg/kg dm oil, treated with hydrophobic nutrients and aerated. Taking the results from the CFU counts and from the analysis of mineral oil and PAH content into account for soil samples with oil amendment or mineral nutrient addition, it seemed that very little had happened.

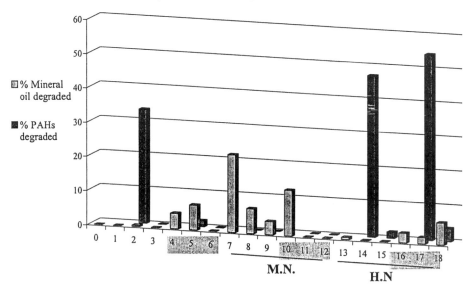

FIGURE 2: Hydrocarbon degradation after 140 days of treatment. Numbers indicate the soil condition (Table 1); indicates the aerated samples.

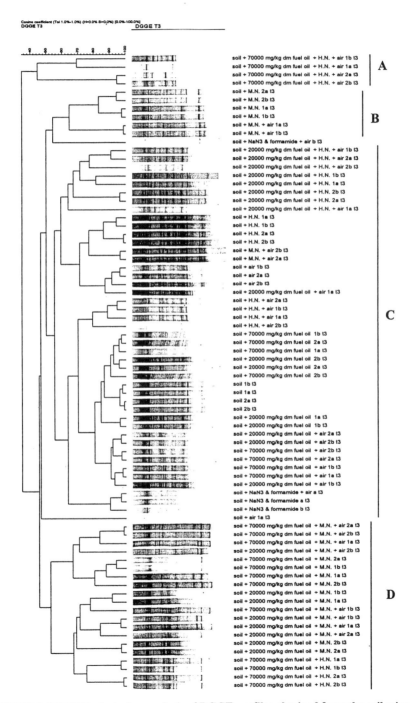

FIGURE 3: UPGMA clustering analysis of DGGE profiles obtained from the soil microcosms after 140 days of treatment (T3). Clustering using the cosine correlation coefficient was performed using Bionumerics (Applied Maths, Belgium) and expressed as percentage similarity. The different conditions are indicated. 1 & 2 represent the two duplicates for each treatment condition; a & b indicate DGGE patterns from 2 samples from the same microcosm.

A low CFU count was observed, and degradation of mineral oil and PAH's seemed negligible (Figure 2). In the case of soil samples with amendment of hydrophobic nutrients, aeration and 70,000 mg/kg dm oil (18a, 18b) a significant increase of oil degrading CFU was noticed. All the colonies displayed the same red morphology. In addition, a decrease of the phenantrene and fluorene content and a small decrease in mineral oil content was observed. Interestingly, these results fit with the results obtained from the PCR-DGGE profile. Indeed, the most distinguished cluster at T3 consisted of the DGGE patterns obtained from the soil samples with 70,000 mg/kg dm oil, hydrophobic nutrients, and aeration. A major increase in the number of oil degraders could also be observed for the non-aerated soil samples amended with 70,000 mg/kg dm oil and treated with hydrophobic nutrients together with a decrease of the phenantrene content. The DGGE patterns resulting from the soils treated under these conditions clustered however with the DGGE patterns of soil samples with oil and mineral nutrients amendment. An increase of LNAPL degrading bacteria could also be observed together with a decrease of mineral oil content, for the non-aerated soil samples without extra oil amendment and with mineral nutrient addition. The results from the cluster analysis were in correlation with the CFU counts and decrease of mineral oil content under these conditions, i.e, a new cluster was formed which contained DGGE profiles originating from the soil samples with mineral nutrient amendment, non-aerated and without oil addition. Furthermore, phenantrene and fluorene were degraded for more than 50% in the soil samples with addition of 20,000 mg/kg dm oil and hydrophobic nutrients (14a, 14b; 15a, 15b) and a developing LNAPL degrading population could be observed. Degradation of mineral oil however did not take place yet and the DGGE patterns of the soil samples treated under these conditions clustered together with untreated soil samples.

REFERENCES

Bakken, L.R. 1997. "Culturable and Nonculturable Bacteria in Soil." In Van Elsas, J.D., J.T. Trevors, and E.M.H. Wellington (Eds.), *Modern Soil microbiology*, pp. 47-61, Marcel Dekker Inc., NY.

Bowlen, G.F., and D.S. Kosson. 1995. "In Situ Processes for Bioremediation of BTEX and Petroleum Fuel Products." In Young, L.Y., and C.E. Cerniglia (Eds.), *Microbial Tranformation and Degradation of Toxic Organic Chemicals*, pp 515-542, Wiley-Liss Inc., NY.

El Fantroussi, S., J. Mahillon, H. Naveau and S.N. Agathos. 1997. "Introduction and PCR Detection of *Desulfomonile tiedjei* in Soil Slurry Microcosms." *Biodegradation, 8*: 125-133.

Heuer, H., and K. Smalla. 1997. "Application of Denaturing Gradient Gel Electrophoresis and Temperature Gradient Gel Electrophoresis for Studying Soil

Microbial Communities." In Van Elsas, J.D., J.T. Trevors, and E.M.H. Wellington (Eds.), *Modern Soil microbiology*, pp. 353-373, Marcel Dekker Inc., NY.

Marchesi, J.R., T. Sato, A.J. Weightman, T.A. Marttin, J.C. Fry, S.J. Hiom and W.G. Wade. 1998. "Design and Evaluation of Useful Bacterium-Specific Primers that Amplify Genes Coding for Bacterial 16S rRNA." *Appl. Environ. Microbiol., 64* (2): 795-799.

FIELD RELEASE OF GENETICALLY ENGINEERED BIOLUMINESCENT BIOREPORTERS FOR BIOREMEDIATION PROCESS MONITORING AND CONTROL

Steven Ripp, David E. Nivens, and Gary S. Sayler
Center for Environmental Biotechnology, University of Tennessee, Knoxville, Tennessee 37996

ABSTRACT: *Pseudomonas fluorescens* HK44 is a bioluminescent bioreporter that emits visible light when degrading naphthalene and other polycyclic aromatic hydrocarbons (PAHs). This strain harbors the bioluminescent reporter plasmid pUTK21 which contains a *nahG-luxCDABE* gene fusion in a salicylate inducible operon. In October of 1996, *P. fluorescens* HK44 was introduced into semi-contained subsurface soil lysimeter structures in the first EPA approved environmental release of a genetically engineered microorganism (GEM) for bioremediation purposes. The study was performed to determine if GEMs could be successfully introduced and maintained during a long-term soil bioremediation process and to show that bioluminescent light emission could be utilized as an *in situ* tool for on-line bioremediation process monitoring and control. HK44 population dynamics were tracked using standard plate count procedures while light emissions were measured using instrumentation consisting of fiber optic guides, novel on-line biosensors, and a portable photomultiplier tube. The largest portion of the field release occurred over an approximate two-year period, during which it was shown that hydrocarbon contamination could be correlated with bioluminescence throughout the bioremediation process, illustrating that bioreporter organisms were capable of providing, in real-time, a general assessment of bioremediation effectiveness. Longer term analysis of *P. fluorescens* HK44 population activities during the third year after release (up to 1150 days after initial release) indicated that HK44 cells were still active and could be induced to bioluminesce upon exposure to naphthalene.

INTRODUCTION

Bioremediation relies on the biodegradative potentials of microorganisms for the removal of hazardous chemicals from contaminated sites. To be effective, it is imperative that the relevant microbial populations remain in a viable and active state throughout the bioremediation process. An attractive tool for monitoring localized microbial processes to determine if they are favorable for bioremediation to occur relies on the use of bioreporter organisms that respond to and signal the presence of specific chemical compounds in their environment. In this study, a *P. fluorescens* strain capable of degrading naphthalene was genetically modified to contain a *lux* gene cassette, thereby endowing it with the ability to generate a bioluminescent light signal in response to naphthalene presence, overall bioavailability, and biodegradative capacity (Heitzer et al., 1992; King et al., 1990). The bioreporter strain, designated *P. fluorescens* HK44, was released into a PAH-contaminated field site consisting of six intermediate-scale lysimeters.

Utilizing light-sensing devices such as photomultiplier tubes and fiber optic guides, a rapid, on-site, on-line, and real-time assessment of biodegradative activities could be achieved. The initial phases of this project have been described elsewhere, therefore only a brief overview will be provided here (Ripp et al., 2000). Emphasis will rather be placed on the long-term survival and genetic stability of the HK44 organisms.

Site Description. The lysimeter structures were located at Oak Ridge National Laboratory, Oak Ridge, Tennessee (Cox et al., 2000). Each lysimeter consisted of a corrugated steel culvert, 4 m deep by 2.5 m in diameter, set into concrete foundations 3 m below ground level. Numerous types of instrumentation were incorporated into each lysimeter. Sensors were buried in the soil to continuously monitor temperature and oxygen concentrations, fiber optic guides and a portable photomultiplier tube (PMT) detected bioluminescent signals, and biosensors identified vapor phase volatile hydrocarbons. An air distribution manifold was placed in the bottom of each lysimeter to provide supplementary aeration. Monitoring devices to measure soil oxygen concentrations and temperature levels were also present.

The lysimeters were filled with a stratified bed consisting of a 31 cm layer of gravel, a 61 cm layer of sand, a 92 cm layer of clean soil, a 92 cm layer of treated soil, and a 61 cm cap of clean soil (Huntington loam type; 42% sand, 40% silt, 18% clay, 1.3% organic carbon). The treated layer of soil consisted of either chemically contaminated or clean soil, with or without a *P. fluorescens* HK44 inoculum. Chemically contaminated soil was artificially contaminated with naphthalene at 1000 mg/kg soil and anthracene and phenanthrene each at 100 mg/kg soil. Soil within the treatment layer was added in approximate 10 cm increments, each being compacted to a bulk density of 1300 kg/m^3. Between each 10 cm increment was sprayed 4 L of a *P. fluorescens* HK44 inoculum in saline, producing initial cell densities of approximately 1 x 10^6 cfu/g soil. One lysimeter served as a control and did not receive an HK44 inoculum. Due to a volatile loss of naphthalene in the initially contaminated and aged soil, an additional pollutant mix was added to select lysimeters on day 135. Twenty-four kilograms of naphthalene and 2 kg of anthracene were dissolved in 833 liters of transformer oil (Exxon Univolt 60), from which 208 liters was applied to each of the selected lysimeters. This produced an approximate loading capacity of 1000 mg/kg throughout the treatment zone. Seven days later, 190 liters of a minimal nutrient media was also added to serve as a basic nutrient source that, accompanying the oil, stimulated HK44 population growth.

MATERIALS AND METHODS

Bioluminescence Detection. Light monitoring equipment consisted of vapor phase naphthalene biosensors suspended at various depths in the soil and a portable PMT that could be lowered directly into the contaminated soil at any depth or location (Sayler et al., 1999). Using these tools, an on-line indication of localized conditions could be quickly assessed to determine if environmental factors favored bioremediation.

The biosensors were utilized for the detection of soil-borne vapor phase naphthalene and were installed in two of the chemically contaminated lysimeters and one of the non-contaminated lysimeters. The biosensors contained alginate-encapsulated *P. fluorescens* HK44 cells placed into a porous, light-tight stainless steel tube (14 cm long by 2.4 cm diameter) (Heitzer et al., 1994). A fiber optic guide was inserted into each biosensor to transfer bioluminescent signals to a computer-based light monitoring system.

The portable PMT consisted of a light-tight housing containing a photon counting PMT module that could be lowered into the treatment zone to monitor light emanating directly from HK44 cells from within the soil matrix.

Soil Sampling and Analysis. Soil cores traversing through the treatment zone layer were removed from each of the lysimeters at various timepoints over an approximate three year period. HK44 populations were assessed from soil samples by plating on a selective media containing tetracycline (Ripp et al., 2000). Soil contaminant concentrations were determined using a solvent extraction and separation procedure followed by gas chromatograph-mass spectrometer analysis.

RESULTS

Bioluminescent Response from *P. fluorescens* HK44. Both the vapor phase biosensor devices and the portable PMT were successful in monitoring bioluminescence from HK44 cells. Bioluminescent signals from biosensors placed in the chemically contaminated lysimeter approached levels of 120,000 counts/second while biosensors in the non-contaminated lysimeter produced only background levels of approximately 800 counts/second due to basal levels of *lux* gene expression. Bioluminescent responses typically persisted for approximately five days, after which the alginate encapsulation matrix became too dehydrated to adequately support continued cell growth and maintenance and would have to be replaced.

The portable PMT probe detected bioluminescence from HK44 cells directly within the soil matrix. Figure 1 illustrates portable PMT-based light data acquisi-

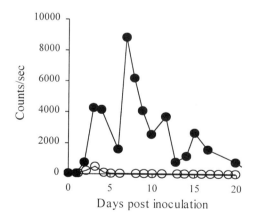

FIGURE 1. Portable PMT derived bioluminescent response from *P. fluorescens* HK44 in chemically contaminated soils (●) and in non-contaminated soils (○). Day 0 corresponds to day 462 of experiment.

tion from within the treatment layer of a chemically contaminated lysimeter as opposed to a non-contaminated lysimeter.

Long-Term Survival and Maintenance of *P. fluorescens* HK44 Populations. The main portion of the field release occurred over a 656 day period, after which spontaneous sampling took place up to a total of 1150 days. Figure 2 illustrates HK44 population dynamics over the complete 1150 day period. Overall, HK44 numbers declined from an initial average of 1.5 ($\pm$ 0.46) x 10^6 cfu/g soil to a final average concentration of 279 $\pm$ 30 cfu/g soil. On day 135, the chemically contaminated lysimeters received a supplementary PAH and inorganic nutrient addition that produced a sharp increase in HK44 numbers up until day 154 when a steady decline continued. Since selective plating was imprecise, due to the presence of microbes besides HK44 capable of growth on the selective media, a quantitative PCR assay was also performed to verify HK44 cell concentrations. Quantitative PCR, using the *luxA* gene as a probe, was performed on samples obtained on day 1150, and indicated that HK44 cells were still present in the lysimeter soils at approximately 1 x 10^5 cfu/g, whether chemically contaminated or not. Granted that this is an estimate that includes living, inactive, and dead cells, it is significantly higher than what was predicted through selective plating. However, it is also assumed that selective plating will underestimate HK44 populations, especially at these late timepoints when a significant portion of the cells may have become nonculturable. Therefore, we can only predict that HK44 populations were present at levels from 1 x 10^2 to 1 x 10^5 cfu/g soil. Although this is a considerable range of values, it must be emphasized that after three years in the soil, the HK44 populations had not become extinct. Additionally, the *lux* insertions, although plasmid-based and void of selective pressures, had remained remarkably stable over time. More interestingly, however, was the fact that HK44 isolates obtained on day 1150 were still capable of generating bioluminescence when exposed to naphthalene vapor.

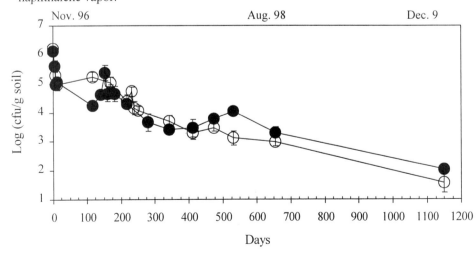

FIGURE 2. Population dynamics of *P. fluorescens* HK44 in the chemically contaminated (●) and non-contaminated (○) lysimeters.

CONCLUSIONS

Bioremediation can be an effective strategy in environmental clean-up efforts if it can be adapted to simple and cost-effective monitoring methods. Soil remediation processes usually employ gas chromatography/mass spectrometry-based detection protocols that are typically time consuming, expensive and requisite of trained personnel. Additionally, results cannot be obtained in real-time or on-line. The results of this study have demonstrated that the implementation of bioreporter organisms into the bioremediation monitoring program can serve as efficient and effective on-line, real-time sensors for contaminant bioavailability. A GEM such as *P. fluorescens* HK44 can bioluminescence to indicate whether conditions are favorable or unfavorable for bioremediation to occur, and, if deemed unfavorable, analytical and microbiological assays and localized treatments can be implemented to diagnose and correct the problem. However, for this strategy to be effective, the GEM must survive throughout the bioremediation process. It is often assumed that GEMs will gradually become extinct after release because they are unable to effectively compete with the indigenous microbial consortia. This reduced level of fitness is thought to be due to the extra energy demands mandated by the presence of the introduced genes and by the fact that GEMs are usually nurtured under optimal laboratory conditions for several years, thus making them physiologically weak and less prone to survive within harsh environments. From this study we have shown that, at least for *P. fluorescens* HK44, survival and persistence are not obstacles to field release.

ACKNOWLEDGMENTS

Research sponsored by the Office of Biological and Environmental Research, U.S. Department of Energy, grant number DE-F602-94ER61870.

REFERENCES

Cox, C. D., D. E. Nivens, S. Ripp, M. M. Wong, A. Palumbo, R. S. Burlage, and G. S. Sayler. 2000. "An intermediate-scale lysimeter facility for subsurface bioremediation research." *Bioremediation 4*: 69-79.

Heitzer, A., O. F. Webb, J. E. Thonnard, and G. S. Sayler. 1992. "Specific and quantitative assessment of naphthalene and salicylate bioavailability by using a bioluminescent catabolic reporter bacterium." *Appl. Environ. Microbiol. 58*: 1839-1846.

Heitzer, A., K. Malachowsky, J. E. Thonnard, P. R. Bienkowski, D. C. White, and G. S. Sayler. 1994. "Optical biosensor for environmental on-line monitoring of naphthalene and salicylate bioavailability with an immobilized bioluminescent catabolic reporter bacterium." *Appl. Environ. Microbiol. 60*: 1487-1494.

King, J. M. H., P. M. DiGrazia, B. Applegate, R. Burlage, J. Sanseverino, P. Dunbar, F. Larimer, and G. S. Sayler. 1990. "Rapid, sensitive bioluminescence reporter technology for naphthalene exposure and biodegradation." *Science 249*: 778-781.

Ripp, S., D. E. Nivens, Y. Ahn, C. Werner, J. Jarrel, J. P. Easter, C. D. Cox, R. S. Burlage, and G. S. Sayler. 2000. "Controlled field release of a bioluminescent genetically engineered microorganism for bioremediation process monitoring and control." *Environ. Sci. Technol. 34*: 846-853.

Sayler, G. S., C. D. Cox, R. Burlage, S. Ripp, D. E. Nivens, C. Werner, Y. Ahn, and U. Matrubutham. 1999. "Field application of a genetically engineered microorganism for polycyclic aromatic hydrocarbon bioremediation process monitoring and control." In R. Fass, Y. Flashner, and S. Reuveny (Eds.), *Novel Approaches for Bioremediation of Organic Pollution*, pp. 241-254. Kluwer Academic/Plenum Publishers, New York.

RAPID, SENSITIVE, AND ACCURATE MONITORING METHOD FOR AUGMENTED DIOXIN-DEGRADING BACTERIA

Hideaki Nojiri, Jaka Widada, Takako Yoshida, Hiroshi Habe, and Toshio Omori (Biotechnology Research Center, The University of Tokyo, Japan)

ABSTRACT: The fluorogenic probes assay, competitive polymerase chain reaction (PCR) and co-extraction of standard cells were combined to develop the rapid, sensitive, and accurate quantification method for the copy number of target carbazole 1,9a-dioxygenase gene (*carAa*) and the cell numbers of *Pseudomonas* sp. strain CA10. The internal standard DNA having the replacement by 20-bp-long DNA region for binding of specific probe in fluorogenic PCR (TaqMan) was constructed, and resultant DNA fragment is similar to the corresponding region of intact *carAa* gene in term of G+C content. This internal standard DNA was used as a competitor in PCR reaction, and could be distinguished from target *carAa* gene by two specific fluorogenic probes with different fluorescence labels, and was automatically detected in single tube using the ABI7700 sequence detection system. Moreover, to overcome variations in the efficiency of cell lysis and DNA extraction between the samples, the co-extraction method was developed. A mini-transposon was used to introduce competitor DNA into the genome of other pseudomonad, and the resultant mutant was used as a standard cell. Using this developed method, the copy numbers of *carAa* gene and the cell numbers of strain CA10 in soil samples were quantified rapidly, precisely, and independently of soil types.

INTRODUCTION

Pseudomonas sp. strain CA10 is revealed to have a high potential to degrade several polyaromatic compounds including dioxins (Nojiri et al., 1999). Successful bioremediation of dioxin-contaminated environment using dioxins-degrading bacterium, strain CA10, requires that this bacterium can survive and have a high catabolic activity to degrade dioxins. Therefore, it is desirable to establish the evaluation methods for monitoring and quantification of strain CA10 in environmental samples.

Competitive PCR is a standard method for quantification of specific DNA sequence in environment samples (Moller and Jansson, 1997; Mesarch et al., 2000). Existing methods for quantification by competitive PCR amplification have been labor intensive, requiring each sample to be separated by gel electrophoresis, followed by quantification of bands after hybridization (Mesarch et al., 2000), or isotopic labeling of the PCR products (Moller and Jansson, 1997). The optimum number of PCR cycles must also be verified before application of these procedures for quantification.

Recently, quantitative PCR using fluorogenic probes assay (TaqMan) has been shown to be a rapid and sensitive method for quantification of PCR products independently of PCR cycles. This system uses a fluorogenic probe, and the amount of fluorescence detected is proportional to the amount of accumulated PCR product (Gibson et al., 1996).

On the other hand, the efficiency of cell lysis and DNA extraction were known to vary for different sample type and DNA extraction procedure (Miller et al., 1999). Therefore, the variation in the efficiency of cell lysis and DNA extraction must be taken into account in order to obtain accurate and reproducible results.

In this study, we describe the development of a novel real-time and accurate method for quantitative determination of *carAa* gene and the cell numbers of strain CA10 in soil samples by a combination of real-time competitive PCR using fluorogenic probes. To overcome the bias in lysis and extraction of DNA from different soil types, co-extraction of standard cell method was also developed.

MATERIALS AND METHODS

Bacterial strains and growth conditions. *Escherichia coli* JM109 was used as the host strain for all plasmids. The carbazole-utilizing bacterium, *Pseudomonas* sp. strain CA10 (Ouchiyama et al., 1993), was used as a test strain to validate the novel competitive PCR method. *Pseudomonas* sp. strain DS1 is not carbazole-degrading bacterium that was isolated as a dimethyl sulfide-utilizing bacterium (Endo et al. unpublished data). *Pseudomonas* sp. strain DS1::*IS* is a strain that was chromosomally tagged with the internal standard DNA as described below. All strains were grown in Luria Bertani broth (LB) supplemented with the following antibiotics when appropriate ampicillin (100 μg ml^{-1}) and/or kanamycin (50 μg ml^{-1}), at 37°C for *E. coli* strains, and at 30°C for pseudomonads.

Oligonucleotide primers and probes. Two specific primers, CarAa3 [position in nucleotide sequence of *car* gene cluster (DDBJ/EMBL/GenBank accession no. D89064: 1,467 to 1,486)] and CarAa2 (position: 1,735 to 1,754), were selected to amplify a 288-bp DNA fragment of the carbazole 1,9a-dioxygenase gene. The annealing site of hybridization probe corresponds to the position 1,699 to 1,719 of the nucleotide sequence of *car* gene cluster. The nucleotide sequences for the oligonucleotide primers and hybridization probes used are listed in Table 1. The standard probe was labeled with the fluorescent VIC (PE Biosystem) on the 5' end and N,N,N',N'-tetramethyl-6-carboxyrhodamine (TAMRA) on the 3' end. The target probe was labeled with the fluorescent 5-carboxyfluoroscein (FAM) on the 5' end and TAMRA on the 3' end. The internal standard probe and the target DNA probe were constructed by PE Biosystems (Ciba, Japan).

DNA manipulations. All DNA manipulations were performed as described by Sambrook et al. (1989). Plasmid DNA was isolated from *E. coli* host cells with the Quantum Prep Plasmid Miniprep Kit (Bio-Rad Laboratories, Richmond, Calif.).

DNA was extracted from agarose gels with Concert™ Rapid Gel Extraction System (Gibco BRL Life Technologies, Madison, WI.), according to the manufacturer's instructions. Restriction enzymes and the DNA ligation kit were purchased from Takara Shuzo Co., Ltd. (Kyoto, Japan).

Construction of standard DNA and verification of amplification efficiency. Two primers, ISF and ISR, were selected to generate a competitive standard differing from the *carAa* sequence by 20-bp. This unrelated 20-bp-long sequence and their complement were underlined in ISF and ISR sequences as shown in Table 2, which contained *Bam*HI and *Xba*I sites. Two separate amplifications were carried out by using two primer sets, CarAa3/ISR and ISF/CarAa2, using plasmid pUCA1 (Sato et al., 1997) as a template. The two amplification products, containing the *carAa* fragment and a single 20-bp overlapping region, were purified from agarose gel, mixed together, annealed and amplified with an out side primer set (CarAa3 and CarAa2) to obtain a 288-bp DNA fragment. These 20-bp long different sequences were used as specific sites for the binding of TaqMan probes. For PCR amplification in these constructions, AmpliTaq Gold (PE Biosystems) was used according to the manufacturer's recommendations, by using GeneAmp PCR System 9600 thermal cycler (PE Biosystems). The PCR program used for all amplification was as follows: 10 min at 96°C; 35 cycles, with 1 cycles consisting of 1 min at 95°C, 1 min at 55°C, and 1.5 min at 72°C; and the last cycle ended by final extension of the products for 10 min at for 72°C. As the target DNA, 288-bp intact DNA fragment was similarly amplified by using the primer set of CarAa3 and CarAa2. The PCR products of competitive standard and the DNA target were cloned into pT7blue(R) vector (Novagen, Madison, WI.), to give recombinant plasmids, pTIS and pTTG, respectively. To clarify the construction, *Bam*HI digestion of PCR products and sequencing of the insert DNA of the plasmids were carried out using PRISM Ready Reaction Terminator Cycle Sequencing Kit (PE Biosystems).

Table 1. Sequences of primers and TaqMan probes used in this study

Primers/Probes	Sequence (5'→ 3')	Tm (°C)
Primers		
CarAa3	CTC-TTG-GCA-AAC-CAT-GTG-CC	60
CarAa2	TAC-TTA-GCC-CGA-CTA-CCG-AC	60
ISF	CAC-TCT-AGA-CGG-ATC-CAC-GCG-GGC-TGA-AAG-AAT-TTG-TCG-G	81
ISR	GCG-TGG-ATC-CGT-CTA-GAG-TGA-ATA-CCC-TGA-TTG-TGT-TCG-C	80
TaqMan probes		
Target probe	Fam-CAG-ACC-CAA-GCG-CAC-GTT-TC-Tamra	63
Standard probe	Vic-CAC-TCT-AGA-CGG-ATC-CAC-GC-Tamra	63

The efficiency of amplification was assessed in parallel with target DNA (linear plasmid of pTTG digested with *Eco*RI) and competitive internal standard DNA (linear plasmid of pTIS digested with *Eco*RI). The concentrations of plasmid was estimated using the absorbance at 260 nm. The target DNA and the standard DNA were serially diluted from starting concentration 10^7 copies down to 10 copies per reaction tube. The dilutions then were used as a template in quantitative PCR (TaqMan) assay as described bellow. The competitive PCR standard curve was generated by quantitative PCR assay using the mixture of target DNA, ranging 10 to 10^7 copies in the presence 10^4 copies of internal standard DNA.

Construction of standard cells. The plasmid pTIS was digested with *Kpn*I and *Sal*I to obtain 340-bp fragment containing the internal standard DNA. The purified fragment DNA was cloned in Tn5-plasmid pTnModOKm (Dannis and Zylstra, 1998) at restriction sites *Kpn*I and *Sal*I resulting in recombinant plasmid pTnISKm. Plasmid pTnISKm was used for chromosomal integration of the internal standard DNA into strain DS1 cells by electroporation with the Bio-Rad Gene Pulser system (Bio-Rad Laboratories, Richmond Calif.) set at 2.5 kV. 200A, and 25 µF. Transformants were selected on LB medium containing kanamycin (50 µg ml^{-1}). To confirm the insertion of internal standard DNA, we performed the PCR analysis of total DNA from selected strain DS1::*IS* using primer set CarAa3/CarAa2 and digestion with *Bam*HI.

Quantitative PCR assay and post-PCR analysis. The PCR mixture (20 µl total volume) which consists of primers CarAa3 and CarAa2 (0.5 µM each), TaqMan probes (internal standard probe and/or target probe) each at 100 nM; dATP, dCTP, and dGTP, each at 200 µM, and 400 µM dUTP, 4 mM MgCl$_2$, 0.01 U Uracil DNA glycosylase per µl, 0.025 U of AmpliTaq Gold per µl, and 1x TagMan PCR buffer (PE Biosystems). Amplification and detection were performed with ABI 7700 sequence detection system (PE Biosystem) with the following profile: 1 cycles of 50°C for 2 min, 1 cycles of 95°C for 10 min, and 50 cycles of 94°C for 15 s and 60°C for 1 min. The fluorescent intensity of each dye was measured at every temperature step and cycle during the reaction. The data were analyzed by Sequence Detector version 1.63 software (PE Biosystems).

Preparation of soil sample and extraction/purification of DNA from soil. Two types soil samples taken from Osaka, Japan, (granitic and granitic upland soils) with different total carbon content, 0.16% and 2.52%, respectively, were used in this study. Two hundred fifty microgram of each soil was introduced into 2-ml Ependorf tube and inoculated with strain CA10 (10^3 CFU) cultured in liquid LB medium at 30°C for 12 h. To ensure absolute and accurate quantification of target DNA from soils samples, the fixed amount (10^4 CFU) of standard cells harboring single copy of internal standard DNA (strain DS1::*IS*) prepared similarly to strain CA10, was added to the soils samples, and was incubated at 4°C for over night. On the other hand, the co-extraction with naked standard DNA (10^4 copy numbers) was performed as a comparative study as described by Moller and Jansson (1997).

The direct extraction method by using sodium dodecyl sulfate treatment and high-temperature (68°C) incubation (Zhou et al., 1996) was used to extract the DNA from soil samples. Purification of crude DNA was carried out using a combination of PVPP spin column (Barthelet et al., 1996) and resin column using Wizard™ DNA Clean-Up System (Promega, Madison, WI.). The purified soil DNA was diluted 10 times in sterile distilled water and used for real-time competitive PCR assay.

RESULTS AND DISCUSSION

Competitive internal standard DNA and PCR efficiency. The internal standard DNA fragment is identical to the corresponding region of *carAa* gene except for the 20-bp-long DNA region, which located in the inner portion of 288-bp-long DNA fragment and used for the annealing site for internal standard-specific fluorogenic probe. This 20-bp-long DNA region has same G+C content with the corresponding region of *carAa* gene. Sequence analysis of PCR product also confirmed that the construction of internal standard was correct (data not shown).

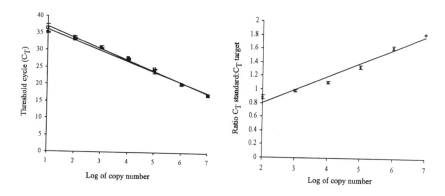

FIGURE. 1. (a) Co-linearity of dilution and assay range of the target DNA (□) and the internal standard of DNA (●), (b) Competitive PCR standard curve. In the single tube, the fixed amount of internal standard DNA (10^4 copies) were co-amplified with 10 fold serial dilution of target DNA.

The amplification plots [copies number versus threshold cycles (C_T)] of target DNA and standard DNA were linear over 7-order magnitude (10 to 10^7 copies per reaction mixture), with the R^2 values of lines being greater than 0.98 in each case as shown in Fig. 1a. These two curves were parallel and overlapped each other, and we could not observed any significant variability in these results, indicating the dynamic range and amplification efficiency of the target DNA and internal standard DNA were quite similar.

The competitive PCR combined with real-time detection was then performed using target DNA, ranging from 10 to 10^7 copies in the presence of 10^4 copies of naked internal standard DNA in single tube. The ratio of the C_Ts of the internal standard and the target DNA was plotted against the copy number of initial target DNA was shown in Fig. 1b. A linear relationship ($R^2 > 0.97$) was observed between 100 copies and 10^7 copies of the target DNA. Unfortunately, we could not observed the linear relationship when 10 copies of target DNA was added to the tube. This result suggests that the detection limit of the competitive PCR in this study was 100-copy numbers of target DNA per tube. To increase the sensitivity up to 1 or 10 copy numbers of target DNA; a combination with the nested PCR is currently being developed.

To be amplified with the same efficiency, the internal standard must share the same primer recognition sites and be as similar as possible to the target sequence in term of G+C content and length (Alexandre et al., 1998). Commonly, competitor DNA has been distinguished from target DNA based on size difference of PCR product. However, the difference in size between the internal standard DNA and target DNA might also affect amplification efficiency. To overcome this problem, an internal standard sharing the same primer recognition sites and similar target sequence in terms of G+C content, and length. A similar approach previously taken by. Alexandre et al. (1998), who used a sandwich hybridization assay was used to quantify the PCR products. However, their approach was more labor intensive, and required post hybridization after PCR amplification.

Competitive internal standard cells. A mini transposon was constructed for stable and single integration of the internal standard DNA into *Pseudomonas* strain DS1. The integration of the internal standard DNA was confirmed by PCR analysis of total DNA strain DS1::*IS* using primer set CarAa3/CarAa2 and digestion with *Bam*HI. The PCR product amplified from strain DS1::*IS* had same size with that from strain CA10 (288-bp) amplified using the same primer set. The digestion of PCR product amplified from strain DS1::*IS* with *Bam*HI gave the similar result with the PCR product of internal standard DNA that was digested with *Bam*HI. These results indicated that the strain DS1:*IS* harboring the internal standard DNA. The insertion of the internal standard DNA was also stable in strain DS1::*IS*, because there was no loss of kanamycin resistance after repeated transfers into nonselective LB medium (approximately 30 generations).

Comparison of co-extraction with standard cells and naked standard DNA. We investigated the accuracy of our developed method using two different types of soil samples in comparison with the previously developed method (Moller and Jansson, 1997). When total DNA was co-extracted with internal standard cells, the copy number of *carAa* gene quantified by real-time competitive PCR was quite similar (Table 2). The difference in copy number of *carAa* gene detected after the co-extraction with standard cells was lower than that detected after the co-extraction with naked standard DNA. This result suggests that, by using the real-time competitive PCR with co-extraction of standard cells, the quantification result is nearly independent on soil types.

The ideal PCR quantification assay should also include a correction factor for differences in lysis efficiency, because there is a possibility that cell lysis is not complete. Previously, Moller and Jansson (1997) have developed a competitive PCR assay that was combined with co-extraction with internal standard naked DNA before DNA extraction, to overcome variation in extraction efficiency. However, we can not exclude the possibility that the quantitative result will be lower than actual cell/copy number in the sample because of incomplete cell lysis. Therefore, the quantification by the Moller and Jansson (1997) method might underestimate the target DNA. Competitive PCR using template DNA obtained by co-extraction with standard cells mostly gave a higher copy number of *carAa* gene than that determined using template DNA obtained by co-extraction with the internal standard DNA (Table 2). In addition, the copy number of *carAa* gene in both soils obtained by co-extraction with standard cell was 1.5-2.5 times higher than the cells number of strain CA10 that was added. We have been reported that strain CA10 has 2 copies of *carAa* gene (Sato et al., 1997), it is considered that almost all *carAa* gene was detected and quantified by this developed method. Based on these results, it was revealed that a marriage between the co-extraction with standard cell and the real-time competitive PCR using fluoregenic probes, has provided a robust, accurate, and rapid quantification of *carAa* gene in natural samples.

Table 2. Quantification of *carAa* gene by real-time PCR combined with co-extraction with internal standard cells or with standard DNA in two soil samples

Co-extraction with	Granitic soil		Granitic upland soil	
	CA10 added (CFU)	*carAa* gene detected (copies)	CA10 added (CFU)	*carAa* gene detected (copies)
standard cells	1.0×10^3	$1.7 \pm 0.80 \times 10^3$	1.0×10^3	$1.5 \pm 0.05 \times 10^3$
naked standard DNA	1.0×10^3	$7.0 \pm 5.20 \times 10^2$	1.0×10^3	$1.3 \pm 0.27 \times 10^3$

The quantification of *carAa* in natural environments will provide useful information to predict their potential on dioxins degradation, due to *carAa* gene involved in dioxins degradation. Moreover, addition of the standard cells before extraction of DNA makes it possible to overcome the variation of efficiency of DNA extraction and cells lysis. Therefore, this approach should be applicable for the quantification of various genes in environmental samples, and also will provide useful and accurate method for evaluation and optimization of DNA extraction and purification methods for environmental sample analysis.

REFERENCES

Alexandre, I., N. Zammatteo, I. Ernest, J.-M. Ladriere, L. Le, S. Hamels, N. Chandelier, B. Vipond, and J. Remacle. 1998. "Quantitative determination of CMV DNA using a combination of competitive PCR amplification and sandwich hybridization." Biotechniques. 25: 676-683.

Barthelet, M., L.G. Whyte, and C.W. Greer. 1996. "Rapid direct extraction of DNA from soil for PCR analysis using polyvinylpolypyrrolidone spin columns." FEMS Microbiol. Lett. 138:17-22.

Dennis, J.J. and G.J. Zylstra. 1998. "Plasposon: Modular self-cloning minitransposon derivatives for rapid genetic analysis of Gram-negative bacterial genomes." Appl. Environ. Microbiol. 64: 2710-2715.

Gibson, U.E., C.A. Heid, and P.M. Williams. 1996. "A novel method for real time quantitative RT-PCR." Genome Res. 6: 995-1001.

Mesarch MB, C.H. Nakatsu, and L. Nies. 2000. "Development of catechol 2,3-dioxygenase-specific primers for monitoring bioremediation by competitive quantitative PCR." Appl. Environ. Microbiol. 66: 678-83.

Miller, D.N., J.E. Bryant, E.L. Madsen, and W.C. Ghiorse. 1999. "Evaluation and optimization of DNA extraction and purification procedures from soil and sediment samples." Appl. Environ. Microbiol. 65: 4715-4724.

Moller, A. and J.K. Jansson. 1997. "Quantification of genetically tagged cyanobacteria in Baltic Sea sediment by competitive PCR." Biotechniques. 22: 512-518.

Nojiri, H., J.-W. Nam, M. Kosaka, K. Morii, K. Furihata, H. Yamane, and T. Omori. 1999. "Diverse oxygenation catalized by carbazole 1,9a-dioxygenase from *Pseudomonas* sp. strain CA10." J. Bacteriol. 181: 3105-3111.

Ouchiyama, N., Y. Zhang, T. Omori, and T. Kodama. 1993. "Biodegradation of carbazole by *Pseudomonas* spp. CA06 and CA10." Biosci. Biotechol. Biochem. 57: 455-460.

Sambrook, J., E.F. Fritsch, and T. Maniatis. 1989. *Molecular cloning: a laboratory manual*. 2 nd ed. Cold Spring Harbor Laboratory Press, Cold Spring Harbor, N.Y.

Sato, S., J.-W. Nam, K. Kasuga, H. Nojiri, H. Yamane, and T. Omori. 1997. "Identification and characterization of gene encoding carbazole 1,9a-dioxygenase in Pseudomonas sp. strain CA10." J. Bacteriol. 179: 4850-4858.

Zhou, J., M.A Bruns, and J.M. Tiedje. 1996. "DNA recovery from soils of diverse composition." Appl. Environ. Microbiol. 62: 316-322.

MONITORING BIOREMEDIATION THROUGH IN-SITU SOIL RESPIRATION

Catherine C. Nestler (Applied Research Associates, Vicksburg MS USA), Lance D. Hansen, (USACOE, Engineering Research and Development Center, Vicksburg MS USA), Scott Waisner (TA Environmental, Vicksburg MS USA), Jeffrey W. Talley (USACOE, Engineering Research and Development Center, Vicksburg MS USA)

Abstract: A device to measure soil respiration in support of landfarm bioremediation was designed and tested at both bench- and pilot-scale. The device, called a soil gas-sampling cell, was assembled, inexpensively, from readily available materials, and could be used with any portable, infrared gas analyzer. The soil used in these experiments was a contaminated soil from a wood treatment facility in El Dorado, Arkansas. Initial contamination levels were 1500 mg/kg pentachlorophenol (PCP) and 13,000 mg/kg of polycyclic aromatic hydrocarbons (PAHs), of which there were 105 mg/kg benzo(a)pyrene (BaP) toxic equivalents. The effect of temperature (long-term and diurnal), nitrogen amendment, and tilling on soil respiration was examined. Soil samples were submitted for phospholipid fatty acid (PLFA) and chemical analysis in order to establish microbial biomass and community changes and correlate these to observed changes in respiration and contaminant concentration. Initial results suggest that soil respiration is a good indicator of microbial community response to oxygen levels (tilling,) nutrient depletion, changes in moisture, and temperature during contaminant degradation. Respiration also reflects changes in the microbial community structure. These results show that in situ respiration measurements could be used to optimize the management of landfarming sites in terms of equipment, personnel, and treatment time.

INTRODUCTION

Microorganisms in the soil consume soil pore oxygen and release carbon dioxide (CO_2) as they grow. This process, respiration, is a segment of the short-term organic carbon cycle in which carbon is converted to CO_2 (mineralization). Several noticeable, ecosystem-level effects can be seen when a soil microbial community is stressed by the presence of a toxin. First, the total community respiration level decreases as the metabolic activity decreases. Then, the respiration to biomass ratio increases. As exposure time to the toxin increases, the nutrient turnover and the rate of organism turnover increases. Microbial community succession demonstrates a decline in species diversity and an overall decrease in growth and respiration with each succession. The most sensitive biomarkers of both soil degradation and repair are those associated with microbial metabolism, such as respiration, biomass, enzyme activity, and/or nitrogen mineralization. While CO_2 production is not a completely accurate assessment of biodegradation of the contaminant, respiration remains the most widely used indirect method to estimate biodegradation in contaminated soils (Zibilske 1994, Li 1998). Soil respiration can be analyzed in the laboratory and in the field by either examining the O_2 consumed or the CO_2 produced. Indirect laboratory measurements, such as electrolytic respirometry, gas chromatography and enzyme analysis, are time-consuming, expensive, require highly trained personnel and are often unsuited for highly contaminated soil. Direct field measurements include alkali trapping of the CO_2, which was the standard field collection method until the development of portable, infrared gas analyzers (IRGA). Studies comparing the two methods have found that IRGA is more suitable for field studies of soil respiration than the traditional alkali method (Bekku et al. 1997).

Bioremediation is an established technology that employs microorganisms to naturally degrade toxic contaminants in soil, sediment and water to less toxic, or non-toxic, forms (USACE, 1996). An inexpensive, adaptable device for rapidly monitoring microbial respiration would allow landfarming management procedures to be continuously modified to maintain an optimum soil environment for microbial growth and, therefore, contaminant degradation.

MATERIALS AND METHODS

The soil used in these experiments was a contaminated soil from a wood treatment facility in El Dorado, Arkansas, described in Hansen et al. (1999). Initial contamination levels were 1500 mg/kg pentachlorophenol (PCP) and 13,000 mg/kg of polycyclic aromatic hydrocarbons (PAHs), of which there were 105 mg/kg benzo(a)pyrene (BaP) toxic equivalents. The experimental design used to test the response of soil respiration to landfarming operational variables at bench and pilot-scale is outlined in Table 1. Soil moisture was maintained at 30-80% of field moisture capacity (FMC) in all experiments.

TABLE 1. Summary of the experimental design used to examine soil respiration in bench and pilot-scale experiments.

Test/Condition	Bench-scale (buckets)	Pilot-scale (LTUs)	Pilot-scale (pans)
Monitoring (minimum)	Daily	2x/week	Weekly/6 months Daily/6 months
Temperature	constant (23°C)	varied	varied
Mixing/tilling	three events	2 week intervals	Weekly
Diurnal cycling	NT	NT	tested
Nutrient addition	NT	Liquid fertilizer, solid fertilizer	Slow-release, solid nitrogen

NT= not tested at this level

The buckets used in the bench-scale tests held 11.5 kg of untreated soil or 8.2 kg of treated soil. A soil gas probe was inserted into the center of each bucket. The pilot-scale LTUs, each 6.1m x 1.2m x 0.61m (20 x 4 x 2 ft) and divided into 20 equal sampling zones, were built to simulate field conditions for traditional landfarming and natural attenuation (Hansen et al. 2000). A soil gas probe was inserted into the center of each zone. The second pilot-scale study used metal pans 3.0m x 0.9m x 0.6m (10 x 3 x 2 ft) to examine the effects of biostimulation and bioaugmentation of contaminated soil (Nestler et al., 2001, in press). Pan 1 (control) was unamended soil. Pan 2 (biostimulation) was supplemented with rice hulls (bulking agent) and dried blood (nitrogen). Pan 3 was augmented with a confirmed biosurfactant-producing bacteria (*P. aeruginosa* strain 64) carried on vermiculite, as well as the rice hulls and nutrients. Each pan was divided into five sampling zones, with a soil gas probes inserted in the center of each zone.

The design of the soil pore gas-sampling probe is illustrated in Figure 1. The top, constructed from a flat, 6-inch circle of 0.32 cm ($1/8^{th}$ in) PVC, surmounted by a dome and a valve, sits aboveground. The variable length slotted pipe is inserted into the soil. Two, 5.08 cm (2-in) high sections are cut from pipes that are 15.24 and 10.16 cm (6-in and 4-in) diameter. These are attached to the top ring. A 5.08 cm diameter (2 in) circle is cut in the center of the ring to accommodate the dome. The junction between the dome and the ring is caulked with a non-binding sealant. The top of the dome is tapped for a 0.32 cm ($1/8^{th}$ in) NPT fitting which accepts the 3-way stopcock. The length of the probe can be adjusted to the required soil depth. Tubing attached to the gas analyzer has a fitting for the stopcock valve. The pump of the gas analyzer withdraws and analyzes an air sample. The probe remains in the ground between sample events.

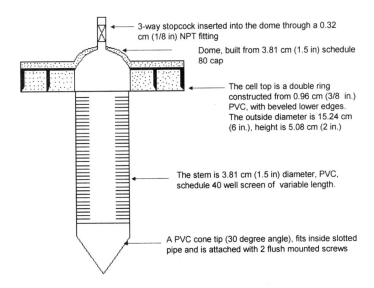

FIGURE 1. Conceptual drawing of the soil gas-sampling cell.

Gas analysis was performed using an LMSx Multigas Analyzer® equipped with an infrared detector and oxygen cell (Columbus Instruments, Columbus, Ohio). Oxygen, carbon dioxide and methane concentrations in the soil were monitored. This unit operates effectively across a temperature range of -10 to $40°$ C. The CO_2 detection range is from 0 to 40% with an accuracy of $\pm\ 0.1\%$. The O_2 detection range is 0 to 25% with an accuracy of $\pm 0.5\%$. The aspiration rate is 100 ml/min. The response time for CO_2 is 20 sec., for O_2, 30 sec.

Microbial biomass and community composition were determined by a modified Bligh-Dyer organic solvent extraction to quantitatively recover bacterial membrane lipid biomarkers (ester-linked phospholipid fatty acids, PLFA) as outlined in White and Ringelberg (1998). Biomass was estimated from the total concentration of membrane lipids and ester-linked PLFA (Balkwill et al. 1988). Contaminant analysis was performed by the Environmental Chemistry Branch of the Waterways Experiment Station on both treated and untreated soil. PAH and PCP concentrations were determined using SW846 EPA Method 8270c for GC/MS after extraction by Method 3540c.

RESULTS AND DISCUSSION

Bench-scale. The respiration in the buckets demonstrated a distinct cycling pattern. After adjusting soil moisture and mixing, the CO_2 concentration dropped and the O_2 concentrations increased. However, within 2 hours, the CO_2 began a rapid increase and the O_2 began to decline. When the soil pore oxygen reached a low of 12-14%, respiration couldn't be sustained and CO_2 production began to decline. The control soil showed a 60-day lag in respiration.

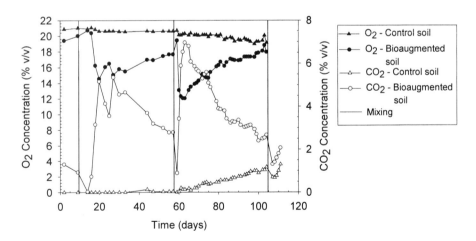

FIGURE 2. Bench-scale in situ respiration of untreated and treated PAH contaminated soil.

Pilot-scale studies. LTUs. LTU 1 was the control unit, representing natural attenuation of the contaminated soil. LTU 2 represented traditional landfarming management. LTU 1 was cultivated only twice, at Day 0, to homogenize the soil, and on Day 71. LTU 2 was tilled every two weeks. Soil gases of both units were sampled twice each week. A cyclic respiration pattern is only somewhat apparent in LTU 2 at this frequency. Cultivation produced a response in both LTUs in the form of increased CO_2 production. A response was also seen after the liquid fertilizer addition on Day 108. A smaller response was observed from solid fertilizer addition on Day 156. In LTU 1, this was probably because the pellets were raked lightly into the surface and not tilled deeply into the soil. The CO_2 production began to increase after 120 days, probably due to acclimation of the microbial communities, since this coincided with the onset of colder weather. In addition to higher respiration levels, LTU 2 also showed greater increases in biomass than LTU 1 and faster contaminant degradation. Another interesting result of the 6-month treatment was the divergence in community composition between the LTUs, although this was not evident from the respiration analysis.

Pilot-scale studies. Pans. Pan 2 is used in this paper as representative of both of the treated soils. Overall soil respiration is shown in Figure 3. The gas concentrations are lower than was seen in the buckets because of the greater exposed surface area in the pans. The respiration response of the control soil to tilling (Pan 1) was minimal to none. The effect of tilling on respiration is highlighted in Figure 4. The O_2-CO_2 decrease/increase pattern first seen in the buckets is also evident at this larger scale. The CO_2 response after tilling was evident by 2-hours.

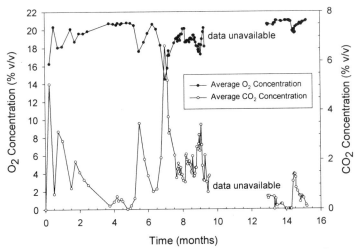

FIGURE 3. Respiration of treated contaminated soil during the study.

None of the pans demonstrated a diurnal pattern. Respiration continued in the treated pans (2 and 3) during the winter months (study months 5-8) when the soil temperature averaged 12°C. Microbial contaminant degradation also continued at high levels during this time (Figure 5). No significant difference (p=0.05) was observed between the final PAH concentrations of Pan 2 and Pan 3, discussed in Hansen et al. (2001, in press). Viable biomass increased in the two treated pans, but not in the control pan, over the first 10 months (Figure 6) which also correlates well with both the respiration and removal data from this time period.

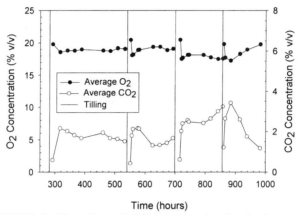

FIGURE 4. The effect of tilling on respiration in the treated soil.

Biomass even increased during the winter. Both treated pans had significant populations of *Pseudomonas* sp. which were not seen in the control pan. When respiration in both experimental pans decreased after 10 months despite moisture adjustment, tilling and optimal temperatures, the soil was checked for nutrients and found to be sub-optimum. The addition of nitrogen resulted in an immediate increase in respiration.

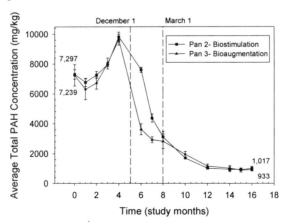

Figure 5. PAH biodegradation in the treated pans during the 16-month study. Winter months of continued degradation are indicated on the graph.

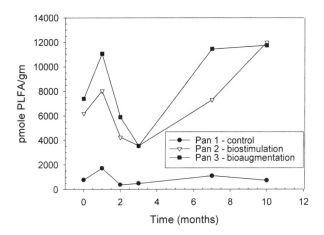

FIGURE 6. Increases in microbial biomass in the treated (Pans 2 and 3) and untreated (Pan 1) soil during the initial 10 months.

CONCLUSIONS

Bioremediation operation and maintenance (O&M) decisions are made based on the environmental requirements of the microbial communities in the soil. It is traditionally believed that, for landfarming treatment, cultivation of the contaminated soil through intensive tilling, and an exhaustive schedule of monitoring and analysis is required for successful bioremediation (USACE, 1996). Pilot-scale studies are recommended for the optimization of field remediation conditions, but the cost of such studies is a primary drawback to their implementation.

While we believe pilot-scale studies to be both necessary and useful in establishing the engineering parameters of field remediation, the soil pore gas probe was developed in an effort to by-pass the need for expensive, multiple year studies and trim on-site remediation costs. The probe allows direct monitoring, via their respiration, of the health of the microbial communities in the soil under treatment. When respiration decreases, the cause can be ascertained. Reasons could be as varied as moisture, oxygen, or nutrient deficit, temperature, or the buildup of toxic intermediates. However, when the cause is known, the appropriate action can be taken. This sequence of events negates the current "shotgun" approach to landfarming O&M in which moisture addition, tilling, sampling and analysis are performed "per schedule" instead of "as needed". Implementing an "as needed" schedule would eliminate both unnecessary chemical and physical analysis of the soil and unnecessary site and equipment maintenance, thus reducing labor and analysis costs associated with site O&M. In addition, while a reduction in tilling frequency reduces costs, it also reduces the possibility of air pollution through the production of fugitive dust, and the possibility of increasing the leaching of the contaminant into groundwater. The soil gas probe is constructed of inexpensive, readily available materials, making it attractive in low-technology situations. It is also adjustable to varying soil depths and relatively inert to most soil contaminants making it applicable to a variety of soil bioremediation scenarios.

Benefits of soil gas monitoring over standard soil sampling/chemical analysis include the ease of sampling, the faster response time, the lower cost involved, and the lack of an effect from soil and contaminant heterogeneity. Because the soil gas monitoring is relatively inexpensive, operation and maintenance costs could be reduced and site operation time decreased.

REFERENCES

Balkwill, D.L., Leach, F.R., Wilson, J.T., McNabb, J.F. and White, D.C. 1988. Equivalence of microbial biomass measures based on membrane lipid and cell wall components, adenosine triphosphate, and direct counts in subsurface aquifer sediments. Microbial. Ecol. 16: 73-84.

Bekku, Y., Koizumi, H., Oikawa, T. and Iwaki, H. 1997. Examination of four methods for measuring soil respiration. Appl.Soil Ecol.5: 247-254.

Hansen, L., Nestler, C., Channell, M., Ringelberg, D., Fredrickson, H. and Waisner, S. 1999. Bioremediation treatability study for remedial action at Popile, Inc., site, El Dorado, Arkansas. Phase II. Pilot-scale evaluation. ERDC/EL TR-00-08, U.S. Army Engineer Research and Development Center, Vicksburg, MS.

Hansen, L.D., Nestler, C., and Ringelberg, D. 2000. "Bioremediation of PAH/PCP Contaminated Soils from POPILE Wood Treatment Facility". In G.B. Wickramanayake, A.R. Gavaskar, J.T. Gibbs and J.L. Means (Eds.), *Proceedings of the Second International Conference on Remediation of Chlorinated and Recalcitrant Compounds*, pp.145-152, Battelle Press, Columbus, OH.

Li, Dong. 1998. Soil gas sampling useful in designing and monitoring of in situ bioremediation processes. Hazardous Waste Consultant 16(2): 1.20-1.22.

Nestler, C., Hansen, L.D., Ringelberg,D. and Talley, J.W. 2001. "Remediation of Soil PAH: Comparison of Biostimulation and Bioaugmentation". In V.S. Magyar, A. Leeson (Eds.), *Proceedings of the Sixth International Symposium on In Situ and On-Site Bioremediation*, Battelle Press, Columbus, OH, in press.

U.S. Army Corps of Engineers. 1996. Bioremediation using landfarming systems. ETL 1110-1-176, 158 pgs.

White, D.C. and D.B. Ringelberg. 1998. "Signature lipid biomarker analysis." In R.S. Burlage, R. Atlas, D. Stahl, G. Geesey, and G. Sayler (Eds.) *Techniques in Microbial Ecology*, pp. 255-272. Oxford University Press, Inc., New York.

Zibilske, L.M. 1994. Carbon Mineralization, In "Methods of Soil Analysis, Part 2. Microbiological and Biochemical Properties", R.W.Weaver et al.(ed). Soil Science Society of America, Madison, WI, p.835-863.

ASSESSMENT OF ALTERNATIVE ENDPOINTS IN LANDFARMING SYSTEMS FOR SUSTAINABLE SOIL USE

Karl C. Nieman (Utah Water Research Laboratory, Utah State Univ., Logan, UT)
Ronald C. Sims (Utah Water Research Laboratory, Utah State Univ., Logan, UT)
Hoi-Ying N. Holman (Lawrence Berkeley National Laboratory, Berkeley, CA)

ABSTRACT: This paper discusses approaches to the elucidation of the chemical nature of bound residues formed by pyrene degradation in a landfarming system at the Champion International Superfund site, Libby, Montana. These include the use of Synchrotron-based FTIR spectromicroscopy to study degradation reactions of pyrene on surfaces in a model vadose zone environment. ^{13}C-NMR studies of pyrene degradation in soil samples from the Champion site will also be utilized to characterize bound residues. Chemical data from these studies will provide information necessary to assess the use of landfarming as a tool for sustainable soil use.

INTRODUCTION

Polycyclic aromatic hydrocarbons (PAH) are one of the primary contaminant classes of concern at the Champion International Superfund Site in Libby, Montana and many other sites where landfarming has been chosen as the primary treatment technology. Landfarming treatment of contaminated soil at the Libby site results in contaminant treatment mechanisms including mineralization and formation of bound residues (Nieman et al., 1999). As the practice of biological treatment of contaminated soils and sediments has become more common, the acceptability of bound residue formation as a treatment endpoint has become a focus with respect to risk assessment at contaminated sites.
Numerous studies have shown that bound residues are a significant endpoint of biological treatment of PAH and other compounds (Singh and Agarwal, 1992; Sims and Abbott, 1993; Hurst et al., 1996; Bhandari, et al., 1996; Guthrie and Pfaender, 1998, Nieman et al., 1999) and several groups have suggested that this is an effective method of soil decontamination (Berry and Boyd, 1985; Bollag, 1992). While present research supports the conclusion that biological treatment of soils results in both contaminant reduction and soil detoxification (Loehr and Webster, 1997; Huling et al., 1995), the mechanisms of PAH bound residue formation in soil systems are currently not well understood. The objective of the research described is to utilize state of the art analytical tools for the development of new approaches to better understand microbial PAH degradation and bound residue formation in the contaminated Libby soil.

ANALYTICAL METHODS

Current work involves the use of synchrotron based Fourier-transform infrared (FTIR) spectromicroscopy at the Lawrence Berkeley National Laboratory (LBNL) Advanced Light Source (ALS) beamline 1.4.3 to observe pyrene degradation in model environmental systems. This instrument was recently used to monitor bacterial chromium reduction on a basalt surface (Holman et al., 1999). The instrument allows for the collection of infrared spectra (Figure 1) and surface area mapping (Figure 2) of environmental samples at a resolution of 10µm or less. Samples of the PAH pyrene, or pyrene with soil humic acid are inoculated with pyrene degrading bacteria and observed over time. This 'direct' observation of the contaminant-bacterial system results in chemical information about the degradation processes taking place on the sample surface as well as the resultant biologically-based binding.

The synchrotron based FTIR microscope at the LBNL ALS utilizes photon energy from a synchrotron source as a light source for a Nicolet Nic-Plan FTIR microscope operating in the mid-IR region from 10,000 cm^{-1} to 450 cm^{-1}. Since the synchrotron beam is many times brighter than conventional IR sources, the beam may be focused to small diameters with little loss of signal. With the 32x objective, the full-width, half-maximum (FWHM) spot size, integrated over all mid-IR wavelengths, is less than 10 µm. This spot size becomes diffraction limited at longer wave-lengths. When measuring areas with diameters less than 100 µm, the synchrotron provides substantial improvement in signal over the conventional globar source.

RESULTS AND DISCUSSION

Our research team has used synchrotron based FTIR spectromicroscopy to characterize the degradation of pyrene on a magnetite surface by three bacterial strains isolated from creosote and pentachlorophenol contaminated soil at the Champion International Superfund site in Libby, Montana. The three strains were characterized as *Mycobacterium* sp. by grams stain, 16S rRNA, and fatty acid analyses. They were isolated from soil samples from the Libby site, are capable of ^{14}C-pyrene degradation, and were shown to metabolize ^{14}C-pyrene to $^{14}CO_2$ in the absence of other carbon sources.

Molecular Monitoring Techniques and Microbial Enumeration 69

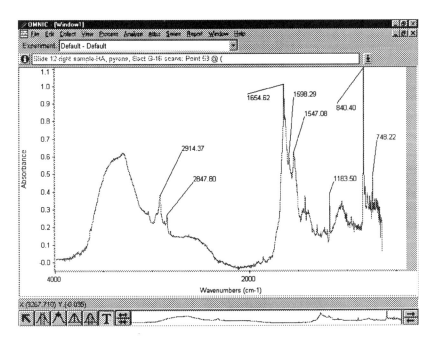

Figure 1. Synchrotron-based FTIR spectromicroscopy spectra of soil humic acid standard after pyrene (peaks at 1183, 840, and 748 cm^{-1}) and pyrene degrading bacteria (peaks at 2914, 2847, and 1654 cm^{-1}) have been applied to the humic acid surface.

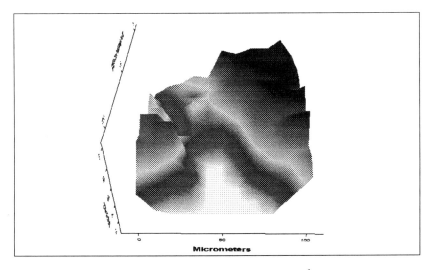

Figure 2. Surface area map of pyrene (peak at 840 cm^{-1}) on sample surface acquired at 10μm resolution

After incubation with pyrene degrading bacteria, samples were solvent extracted and reanalyzed. This allowed for the chemical assessment of the humic surface where bound residues may be present. This 'in situ' chemical analysis can not be accomplished using traditional treatability study techniques that involve the use of chemical extractants or costly radiolabeled compounds. Soil incubations with ^{13}C-pyrene can be analyzed with ^{13}C-NMR and used to provide additional chemical information about the nature of the bound residues produced in the contaminated soil at the Libby site.

Degradation of pyrene on an inoculated humic acid surface is currently being studied. Planned research involves the use of the instrument to monitor degradation of pyrene, and then assess possible changes in the humic acid spectrum due to the potential presence of bound pyrene residues.

The new application of the analytical instrumentation described is providing new tools for evaluating real-time biodegradation and immobilization processes associated with soil organic matter in landfarming treatment systems. An alternative endpoint that includes biologically based binding is especially important for highly hydrophobic target chemicals including the PAHs.

ACKNOWLEDGMENTS

This work was supported by the Huntsman Environmental Research Center at Utah State University, and the U.S. Environmental Protection Agency through the Great Plains/Rocky Mountain Hazardous Substance Research center (Project 93-21).

REFERENCES

Bhandari, A., Novak, J. T., Berry, D.F. (1996). Binding of 4-monochlorophenol to soil. *Environ. Sci, Technol.*, 30, 2305-2311.

Berry, D.F., Boyd, S.A. (1985). Decontamination of soil through enhanced formation of bound residues. *Environ. Sci. Technol.*, 19, 1132-1133

Bollag, J.-M. (1992). Decontaminating soil with enzymes. *Environ. Sci. Tech.*, 26, 1876-1881.

Guthrie, E. A., Pfaender, F. K. (1998). Reduced pyrene bioavailability in microbially active soils. *Environ. Sci. Technol.*, 32, 501-508.

Holman, H.-Y. N., Perry, D. L., Martin, M. C., Lamble, G. M., McKinney, W. R., Hunter-Cevera, J. C. (1999). Real-time characterization of biogeochemical reduction of Cr(VI) on basalt surfaces by SR-FTIR imaging. *Geomicrobiology Journal*, 16:307-324.

Huling, S. G. Pope, D. F. Matthews, J. E. Sims, J. L. Sims, R. C. Sorenson, D. L. (1995). Land Treatment and the Toxicity Response of Soil Contaminated with Wood Preserving Wastes. Remediation, Sprin 1995, 41-55.

Hurst, C. J., Sims, R.C, Sims, J.L., Sorensen, D.L., McLean, J.L., Huling S. (1996). Polycyclic aromatic hydrocarbon biodegradation as a function of oxygen tension in contaminated Soil. *J. Haz Mat. 51*: 193-208.

Loehr, R.C., Webster, M.T. (1997). Changes in toxicity and mobility resulting from bioremediation processes. *Bioremediation Journal*, 1, 149-163.

Nieman, J. K. C., Sims, R.C., Sims, J.L., Sorensen, D.L., McLean, J.E., Rice, J.A. (1999). ^{14}C-pyrene bound residue evaluation using MIBK fractionation method for creosote contaminated soil." *Environ. Sci. Technol.*, 33, 776-781.

Sims, R. C., and C. K. Abbott. 1993. "Evaluation of Mechanisms of Alteration and Humification of PAHs for Water Quality Management." United States Geological Survey report G-1723, USGS, Reston, VA.

Singh, D.K., Agarwal, H. C. (1992). Chemical release and nature of soil-bound DDT residues. *J. Agric. Food Chem.*, 40, 1713-1716.

MICROBIAL HETEROGENEITY IMPLICATIONS FOR BIOREMEDIATION

Susan M. Pfiffner,[1] Anthony V. Palumbo,[2] Barry L. Kinsall,[2] Aaron D. Peacock,[1] David C. White,[1] and Tommy J. Phelps[2]

[1]The University of Tennessee, Knoxville, Tennessee, U.S.A.
[2]Oak Ridge National Laboratory, Oak Ridge, Tennessee, U.S.A.

ABSTRACT: Horizontal heterogeneity with respect to microbial community structure has been rarely examined but could be an important consideration for design of in situ bioremediation. To address this concern, sediment samples were analyzed from four coreholes (C, E, F, and H) at Dover Air Force Base in Dover, Delaware. Three replicate sediment samples were recovered within the 1.7-to 2.0-m interval from each corehole. Sediments were analyzed for grain size, phospholipid composition, and microbial enumerations (colony-forming units and most-probable number dilution series). The phospholipid analysis showed shifts in the Gram-positive and Gram-negative communities; these shifts corresponded with changes in grain size. In addition, microbial counts decreased with the higher clay content of sediments. A hierarchical cluster analysis of the phospholipid microbial community profiles was used to compare and evaluate horizontal heterogeneity in relationship to sample variability. In the vadose zone sediments, the three replicate sediment samples taken within the 1.7- to 2.0-m interval from each corehole clustered together. The community similarity was greatest between samples from coreholes C and E at 30 m apart; communities in these samples clustered together. The most divergent community was observed in corehole F with the highest clay content, which was located 40 m from corehole E and 60 m from corehole H. There was stronger spatial discrimination among communities at the tens-of-meters scale than was evident within a corehole. This discrimination is important because it gives confidence that differences in communities seen at the tens-of-meters scale are not random fluctuations (if this were so, one might expect some of the replicates from different coreholes to cluster). These results provide evidence that the spatial structure of subsurface microbial communities can be dependent on the spatial structure of key physical and chemical properties (e.g., clay content) of the subsurface media.

INTRODUCTION

Microbial heterogeneity occurs at many scales in natural environments. Analyses of microbial heterogeneity often focused on scales ranging from centimeters to meters within a corehole (Balkwill *et al.*, 1989; Brockman *et al.*, 1992). For example, Smith *et al.* (1992) observed 20-fold changes in microbial abundance and activity within a 9-m vertical profile. Similarly, Beloin et al. (1988) reported a 20-fold change within a 20-cm interval. The issue of microbial heterogeneity is important for the assessment, design, and implementation of bioremediation. The goal of this study was to examine horizontal heterogeneity at a scale of tens of meters within a single sedimentary unit of a shallow vadose zone. We utilized both cultivation and non-cultivation microbial assays to provide estimates of microbial abundance and microbial community structure as a means to determine microbial heterogeneity. In addition, physical analyses were used to determine grain size, a parameter that may influence microbial communities. We observed a greater distinction in microbial community structure between sediments that contained increased proportions of fine grain and clay content in contrast to sediments that contained sands. The microbial communities showed a strong horizontal spatial discrimination at the tens-of-meters scale. The results provide evidence that the spatial structure of subsurface microbial communities can be dependent on the spatial structure of key physical and chemical properties (e.g., clay content) of the subsurface media.

MATERIALS AND METHODS
Field Site and Geology
The field site was located between building 428 and the swimming pool within Dover Air Force Base (DAFB) in Dover, Delaware. The geology at DAFB consists of a shallow subsurface that is comprised of the Columbia Formation (Pleistocene), which is −9 m thick in the field site area and overlies the Calvert Formation (Miocene). The Columbia Formation consists of reddish-orange, brown-to-tan sand with minor gravel. The underlying Calvert Formation is 90 m thick and consists predominantly of gray-to-bluish/greenish-gray sandy silt and subordinate shelly sands. Four coreholes were sampled in triplicate at depths of 1.7 to 2.0 m below land surface; these samples were collected at 1.5 m or more above the water table. The coreholes were located in a line from the swimming pool to the far end of building 428. Corehole C was located 30 m away from corehole E, while corehole F was 40 m from corehole E and 70 m from corehole C. Corehole H was 60 m away from corehole F and 130 m from corehole C.

Anaerobic Soil Sampling and Processing
Sediments were collected in lexan liners through rotasonic drilling techniques and were processed under anaerobic conditions to preserve the natural characteristics of the anaerobic microbial community (Pfiffner et al., 1995). Samples were collected in triplicate from the vadose zone within each of the four coreholes, which were designated C, E, F, and H. The triplicate samples were recovered as replicates: replicate A from 1.7 to 1.8 m, replicate B from 1.8 to 1.9 m, and replicate C from 1.9 to 2.0 m. All sediment processing was conducted within an airtight glove box containing an argon gas atmosphere. Only the inner portion of the sediment cores was extracted for physical and microbial analysis. Sediment samples were collected from each 10-cm section by means of sterile utensils and were placed in whirlpak bags. The samples for lipid analysis were immediately frozen with dry ice to preserve the lipid biomass, while the samples for particle size analysis and enumeration studies were placed within a mason jar and sealed while inside the glove bag. The sediment samples were shipped to the university overnight with ice or dry ice for preservation. TCE was not a factor in the vadose

zone because TCE contamination at the DAFB is at or below the water table.

Particle Size Analysis

Vadose zone sediments (15 to 20 g dry weight) were analyzed by particle size analysis. The analysis was conducted by using the dry sieving method for fractionation of sand particles (Gee and Baruder, 1986). This analysis is based of the U.S. Department of Agriculture scheme for particle sizes. The mean diameter of the sample grains was calculated on the basis of weight percentages for all size groups.

Microbial Enumeration Studies

Microbial population densities were determined for the vadose zone sediments (10 g wet weight) by using plate counts and serial dilution with media targeted for different physiological groups that represent aerobic and anaerobic heterotrophs (Pfiffner *et al.*, 1997a, 1997b).

Lipid Extraction and Analysis

Microbial biomass and community structure were assessed by analyzing the lipid composition of the microbial community. Lipid extractions of the samples were performed to determine the polar lipid fatty acid (PLFA) profiles. The frozen sediment samples (75 g dry weight) were lyophilized to remove water because it may interfere with the extraction efficiency. The membrane lipids were extracted by means of a modified buffered chloroform and methanol solvent system (Bligh and Dyer, 1959; White and Ringelberg, 1998). The lipid classes were separated by silicic acid chromatography by means of the micro technique, and the polar lipids were subjected to mild alkaline methanolysis (Tunlid *et al.*, 1989). The PLFAs were identified and quantified through gas chromatography/mass spectroscopy (Nichols *et al.*, 1986; White and Ringelberg, 1998). For preliminary analyses, we grouped the phospholipids into classes (e.g. Table 2), averaged the profiles of the replicates in each corehole, and examined this combined data. Hierarchical cluster analysis was performed on the entire set of phospholipid profiles for each sample (not the combined data) by using Statistica software.

RESULTS AND DISCUSSION

Grain size

According to the grain size analysis, vadose sediments from coreholes C, E, and H mainly consisted of coarse-to-fine grain sands (Table 1). While sediment from corehole F had the largest percentage of very coarse sand, it also had the largest combined percentage of very fine sands, silts, and clays. Grain sizes in these DAFB samples were consistent with grain sizes from the Abbott's Pit site and Oyster, Virginia, site located further south on the Delmarva Peninsula (Zhang *et al.*, 1998). The heterogeneity shown in these samples along a horizontal strike was not expected to be as dramatic as the heterogeneity shown in samples along a depth profile (Zhang *et al.*, 1997).

TABLE 1. Grain Size Averaged Across Replicates (%)						
SAMPLE*	VCS	CS	MED	FINE	VFINE	SILTS	CLAYS
C	13.05	32.52	17.86	21.89	7.31	0.07	0.00
E	15.28	40.11	16.43	18.55	8.92	0.18	0.57
F	28.03	38.24	7.70	14.83	10.11	0.83	1.62
H	13.78	49.68	15.74	14.00	6.53	1.31	0.00

* Vadose sediment samples for each corehole.
VCS = very coarse sand; CS = coarse sand; MED = medium sand;
FINE = fine sand; VFINE = very fine sand.

Microbial enumeration
Aerobic microbial enumeration studies showed that the broth enumerations gave higher values, 4.0×10^2 to 3.4×10^5 cells/g, than did the plate counts, which ranged from 6.6×10^1 to 1.0×10^4 cells/g (data not shown). Corehole H had the highest number of aerobic and anaerobic heterotrophs: 3.4×10^5 and 7.0×10^2, respectively. In contrast, corehole F had the lowest cell numbers for aerobic heterotrophs: 4.0×10^2. However, corehole F had the second highest cell numbers for anaerobic heterotrophs: 3.7×10^2. The lower values for aerobic heterotrophs in the more clayey sediments are similar to results from subsurface sampling at the Savannah River Site (Phelps et al., 1994). The presence and detection of anaerobic bacteria in vadose sediments has been previously documented by Brockman et al. (1992), who demonstrated that these microorganisms exist in microniches. The anaerobic microbial communities, although lower in abundance than aerobic microbial communities, may still play a role in the microbial metabolic and degradative processes.

Lipid Analysis
Phospholipid biomass measurements ranged from 5.53 to 74.30 pmol/g dry weight. This indicates, as expected, that the phospholipid biomass (1.4×10^5 to1.8×10^6 cells/g) was greater than the viable biomass determined by the plate count or broth enumeration assays. This difference in microbial biomass measured by the various methods is typically seen in environmental samples. Based on the phospholipid class groupings (Table 2), the sediments from corehole C and E were very similar. The phospholipid classes are indicative of certain microbial populations: the terminally branched saturated fatty acids represent Gram-positive communities, the monounsaturated and branched monounsaturated fatty acids represent Gram-negative communities, and the mid-chain branched fatty acids represent actinomycete and/or sulfate-reducing communities. Similar in profile to coreholes C and E, corehole H exhibited a slight increase in the proportion of normal and terminally branched saturated fatty acids, a decrease in the proportion of mid-chain branched fatty acids, and the presence of polyunsaturated fatty acids indicative of eukaryotic influences. Corehole F exhibited a starkly different phospholipid profile, with a smaller proportion of monounsaturated and branched monounsaturated fatty acids and a slight increase in the proportion of terminally branched and mid-chain branched fatty acids. The profile for corehole F indicated a decrease in the Gram-negative community and

an increase in the Gram-positive community. This shift toward Gram-positive communities in sediments containing increased proportions of finer grains and clay content follows a trend previously described (Phelps et. al., 1994).

TABLE 2. Phospholipid Composition Averaged Across Replicates (relative percent)

SAMPLE	NSATS	TBSATS	MBSATS	BRMONOS	MONOS	POLYS
C	19.88	23.41	31.63	2.59	22.49	0.00
E	22.88	26.00	27.11	2.14	21.44	0.00
F	26.17	29.55	32.94	0.44	9.34	0.00
H	27.80	27.58	20.30	1.75	20.20	1.85

SAMPLE = vadose sediment samples for each corehole;
Phospholipid classes: NSATS = normal saturates; TBSATS = terminally branched saturates; MBSATS = mid-chain branched saturates; BRMONOS = branched monounsaturates;
MONOS = monounsaturates; POLYS = polyunsaturates.

A more detailed look at the individual phospholipids and ratios of lipids indicated other differences among the samples. Certain phospholipid ratios have been used to define the physiological or nutritional status of microbial communities (Nichols et al., 1986; Pfiffner et al., 1997a; White and Ringelberg 1998). For example, the Gram-negative community in corehole F showed greater signs of stress than the Gram-negative communities in the other three coreholes. The 18-carbon length monounsaturated fatty acid ratio of *trans* to *cis* for corehole F averaged 1.54 in comparison with 0.14–0.21 for the other coreholes. Furthermore, the ratio of cyclopropyl to monounsaturated fatty acids for corehole F was 5.65 in comparison with 0.78–1.67 for the other coreholes. These ratios indicated that the corehole F samples were more likely exposed to environmental stresses such as nutritional limitation. Similar stress responses were seen under starvation conditions, under nutrient-limiting conditions, and under solvent contamination (Kieft et al., 1994; Phelps et al., 1994; Pfiffner et al., 1995). One of the physical factors that may be implicated in causing environmental stress for corehole F is the increased proportion of fine grain sands, silts, and clays which reduces water and nutrient flux in the sediment. The lower microbial abundance observed in the corehole F sediments supports the finding that conditions are nutrient limited in these sediments.

In a hierarchical cluster analysis (Figure 1) of all of the individual fatty acids in the phospholipid profile from the 12 vadose zone sediments, the replicate samples (at 10-cm intervals) from each of the four coreholes clustered together. If random fluctuations among microbial communities had been seen, some but not all of the replicates from different coreholes would have clustered together. This was not the case for these sediments, which clustered within a corehole, suggesting that the community fluctuations were not random. Coreholes C and E, which were located 30 m apart, were most similar in their phospholipid profiles. The community similarity (with a linkage distance of 0.6) was also greatest between these two sites. Corehole H, which was 130 m from Corehole C and 100 m from corehole E, had phospholipid profiles that were less similar to those of corehole C and E and had a linkage distance of 1.25 with respect to profiles from corehole C and E. The most divergent community was observed in corehole F, which was 40 m from corehole E and 60 m from corehole H. Corehole F, with a linkage distance of 2.5, was completely distinct from coreholes C, E, and H. The distinctiveness of corehole F sediments appears to be related to the change in community structure in response to the higher proportion of fine grains, silts, and clay observed in these sediments.

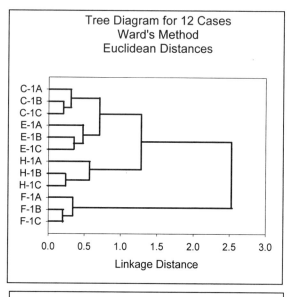

FIGURE 1. Hierarchical cluster analysis of the phospholipid profiles for the vadose zone.

This difference further emphasizes the importance of the changes in microbial community structure with respect to changes in the constituents (i.e., the increase in clay content of the subsurface material). These differences in microbial heterogeneity along a horizontal transect is a condition that should be considered during the design and implementation of a bioremediation effort.

CONCLUSIONS

Microbial communities shifted from Gram-negative to Gram-positive in sediments containing a higher proportion of finer grain sands and clays. In addition, the sediments containing increased clay content exhibited lower microbial abundance, perhaps as a result of less water flux in lower permeability clay sediments. There was stronger spatial discrimination horizontally among communities as defined by PLFA analysis at the tens-of-meters scale, but no spatial discrimination was evident vertically within the vadose zone in each corehole. This observation of horizontal spatial discrimination is supported by the changes observed with the grain size and microbial enumeration analyses. This is important as it gives confidence that differences in communities seen at the tens-of-meters scale are not random fluctuations (if this were so, one might expect the replicates from different coreholes to cluster). These results provide evidence that the spatial structure of subsurface microbial communities can be dependent on the spatial structure of key physical and chemical properties of the subsurface media (e.g., clay content) even within the same sedimentary formation. The data indicates that changes in community structure over tens of meters can be measured and are detectable over background community fluctuations.

Examination of more data and analyses at other sites may statistically validate the implications of spatial microbial heterogeneity in bioremediation.

ACKNOWLEDGMENTS

Special thanks go to Shirley Scarborough, Lisa Fagan, and Chuck Payne, who assisted with the experiments at the University of Tennessee, and to Chunlun Zhang and Mary Ann Bogle, who assisted in field sampling. This research was supported by DOE-Office of Science and . We also thank Tim Griffin and Golder Associates for field site management and sonic drilling operations with Boart Longyear Company. Additional thanks go to Michael Mickula for logistical support at Dover Air Force Base. UT-Battelle LLC manages Oak Ridge National Laboratory for the U.S. Department of Energy under contract number DE-AC05-00OR22725.

REFERENCES

Balkwill, D.L., J.K. Fredrickson, and J.M. Thomas. 1989. Vertical and horizontal variations in the physiological diversity of the aerobic chemoheterotrophic bacterial microflora in deep southeast coastal plain sediments. Appl. Environ. Microbiol. 55:1058-1065.

Beloin, R.M., J.L. Sinclair, and W.C. Ghiorse. 1988. Distribution and activity of microorganisms in subsurface sediments of a pristine study site in Oklahoma. Micro. Ecol. 16:85-97.

Bligh, E.G., and W.J. Dyer. 1959. A rapid method of total lipid extraction and purification. Can. J. Biochem. Physiol. 37:911-917.

Brockman, F.J., T.L. Kieft, J.K. Fredrickson, B.N. Bjornstad, S.W. Li, W Spangenburg, and P.E. Long. 1992. Microbiology of vadose zone paleosols in south-central Washington state. Micro. Ecol. 23:279-301.

Gee, G.W., and J.W. Baruder. 1986. Particle size analysis. In: Klute, A. *et. al.* (ed.) Methods of Soil Analysis, Part 1. Physical and mineralogical methods.

Kieft, T.L., D.B. Ringelberg, and D.C. White. 1994. Changes in ester-linked phospholipid fatty acid profiles of subsurface bacteria during starvation and desiccation in a porous medium. Appl. Environ. Microbiol. 60(9):3292-3299.

Nichols, P.D., J.B. Guckert, and D.C. White. 1986. Determination of monounsaturated fatty acid double-bond position and geometry for microbial monocultures and complex consortia by capillary GC-MS of their dimethyl disulphide adducts. J. Microbiol. Methods 5:49-55.

Pfiffner, S.M., D.B. Ringelberg, D.B. Hedrick, T.J. Phelps, and A.V. Palumbo. 1995. Subsurface microbial communities and degradative capacities during trichloroethylene bioremediation. pp. 263-271. In: Bioremediation of Chlorinated Solvents (Hinchee, R. E., A. Leeson and L. Semprini, eds.), Battelle Press, Richland, WA.

Pfiffner, S.M., A.V. Palumbo, T. Gibson, D.B. Ringelberg, and J.F. McCarthy. 1997a. Relating groundwater and sediment chemistry to microbial characterization at a BTEX-contaminated site. Appl. Biochem. Biotech. 63-65:775-788.

Pfiffner, S.M., J.J. Beauchamp, T.J. Phelps, AV. Palumbo, and T.C. Hazen. 1997b. Effects of methane dosing to groundwaters and sediments. J. Indust. Microbiol. 18 (2/3):204-212.

Phelps, T.J., S.M. Pfiffner, K.A. Sargent, and D.C. White. 1994. Factors influencing the abundance and metabolic capacities of microorganisms in eastern coastal plain sediments. Microb. Ecol. 28(3):351-364.

Smith, R.L., R.W. Harvey, and D.R. LeBlanc. 1991. Importance of closely spaced vertical sampling in delineating chemical and microbiological gradients in groundwater studies.
J. Contam. Hydrol. 7:285-300.

Tunlid, A., D. Ringelberg, C. Low, T.J. Phelps, and D.C. White. 1989. Analysis of phospholipid fatty acids from bacteria at picomolar sensitivities from environmental samples. J. Microbiol. Methods 10:139-153.

White, D.C., and D.B. Ringelberg. 1998. Signature lipid biomarker analysis pp. 255-272. In: Techniques in Microbial Ecology (R.S. Burlage, ed.), Bermedica Product Ltd., Columbia, MD.

Zhang, C., R.M. Lehman, S.M. Pfiffner, S.P. Scarborough, A.V. Palumbo, T.J. Phelps, J.J. Beauchamp, and F.S. Colwell. 1997. Spatial and temporal variation of microbial properties at different scales in shallow subsurface sediments. Appl. Biochem. Biotech. 63-65:797-808.

Zhang, C., A.V. Palumbo, T.J. Phelps, J.J. Beauchamp, F.J. Brockman, C.J. Murray, B.S. Parsons, and D.J.P. Swift. 1998. Grain-size and depth constraints on microbial variability in coastal plain subsurface sediments. Geomicrobiology 15(3):171-185.

DEGRADATION OF PYRIDINE BY *RHODOCOCCUS OPACUS* UFZ B408

Christina Gisela Föllner (Saxon Institute of Applied Biotechnology at the University of Leipzig, Leipzig, Germany)
Wolfgang Babel (UFZ Centre for Environmental Research Leipzig-Halle, Department of Environmental Microbiology, Leipzig, Germany)

ABSTRACT: *Rhodococcus opacus* UFZ B408 is able to grow on pyridine. As a rule the rate of consumption, i.e. detoxification is low. Knowledge on the metabolism of pyridine, regulation and growth rate limiting steps is useful in order to optimise the pyridine degradation by physiological measures or by means of genetic engineering.
First investigations of *Rhodococcus opacus* showed (Brinkmann and Babel, 1997) that with growth on pyridine an $NAD(P)^+$-specific glutaric dialdehyde dehydrogenase is induced, which is considered a key enzyme in the assimilation of pyridine. Therefore, we decided to purify this enzyme in order to study its regulation. The results obtained are presented. The purification procedure involved DEAE-Sepharose column chromatography on CL-6B and Fast-Performance Liquid Chromatography (FPLC), using a Phenyl-Superose HR5/5 column (Fa. Pharmacia LKB Biotechnologie GmbH, Freiburg, Germany). The molecular weight of the partially purified enzyme determined by gel filtration was approximately 200 kDa. The N-terminal amino acid sequence was determined. Two separate oligonucleotides (CGF01 and CGF01.1) were deduced from this sequence and labelled with DIG-dd UTP (Fa. Boehringer Mannheim, Germany). Hybridizing with *Bam*HI digested genomic DNA of *R. opacus* and the DIG-dd UTP marked oligonucleotide probe CGF01 gave one significant hybridization signal with an approximately 3.3-kbp *Bam*HI-DNA fragment. The DIG-dd UTP-labelled oligonucleotide probe was used to screen a *R. opacus Bam*HI genomic library in the cosmid pHC79 by colony hybridization. Five positive clones were detected.

INTRODUCTION
Pyridine is a by-product of coal gasification. It is toxic and environmentally persistant and must be detoxified. Ecosystems polluted with pyridine can be decontaminated by means of microorganisms. A number of bacteria, which are capable of growing on pyridine as sole source of carbon, energy, and nitrogen, have been isolated (Ensing and Rittenberger, 1963; Watson and Cain, 1972; Shukla, 1973; Shukla and Kaul, 1974, 1975, 1986; Korosteleva et al., 1981; Sims et al., 1986; Zefirov et al., 1994; Kaiser et al., 1996; Brinkmann and Babel, 1996, 1997; Uma and Sandhya, 1997; Fetzner, 1998). Several groups have investigated the assimilation

pathway. Watson and Cain (1975) suggested two metabolic routes (Fig. 1). In *Nocardia* sp. (strain Z1) glutaric semialdehyde is an early intermediate of pyridine degradation. The formation of an intermediate with 5 C-atoms indicates that the pyridine ring must be cleaved between the heteroatom and the C-2. Furthermore it was shown that with growth on pyridine both an $NAD(P)^+$-specific glutaric semialdehyde dehydrogenase and an $NAD(P)^+$-specific glutaric dialdehyde dehydrogenase are induced.

The metabolites produced by *Bacillus* strain 4 are similar to those obtained during degradation of pyridine by *Corynebacterium* and *Brevibacterium* spp. Pyridine is degraded to succinic acid semialdehyde, and this intermediate is transformed by an inducible NAD^+-specific succinic acid semialdehyde dehydrogenase into succinic acid. The formation of succinic acid semialdehyde and formic acid from the carbon 2 atom of the pyridine ring indicates that the ring is cleaved between carbons 2 and 3. The metabolic steps in the transformation of pyridine to succinic acid remain somewhat speculative (Fig.1). Knowledge of the metabolic pathway is necessary if people want to improve the efficiency of degradation and accelerate the consumption rate.

Objective. Although a variety of microorganisms are able to utilize pyridine, the rate of productive, growth-associated degradation is low. This is also true with *Rhodococcus opacus* UFZ B408. *R. opacus* seems to express the pathway as described for the assimilation of pyridine by *Nocardia* strain Z (Watson and Cain, 1975). In this sequence $NAD(P)^+$-specific glutaric dialdehyde dehydrogenase is a key enzyme (Brinkmann and Babel, 1997). To see whether or not it is growth rate limiting, studied properties of this enzyme.

MATERIALS AND METHODS
Bacterial strains and media. *Rhodococcus opacus* UFZ B408 was grown at 30°C in mineral salts medium (MM) as described by Brinkmann and Babel (1996). For growth on fructose as the sole source of carbon and energy 3 g NH_4Cl per litre medium was added. Fructose was added to MM up to a concentration of 4 g/l, pyridine up to a concentration of 350 mg/l.

E. coli strains (*E. coli* S17-1, Simon et al., 1983; DH5α, Sambrook et al., 1989), were grown at 37°C in Luria-Bertani (LB) medium. For maintenance of plasmids (pHC79, Hohn and Collins, 1980; and pYES2, Fa. Invitrogen, Netherlands) in *E. coli* and its recombinants, ampicillin (Ap) was added at a concentration of 100 mg/l.

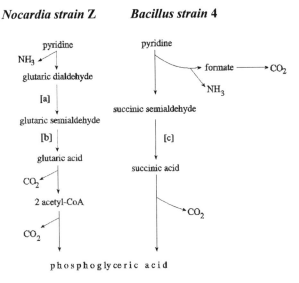

FIGURE 1. Proposed metabolic routes for the metabolism of pyridine (Watson and Cain, 1975); [a] NAD(P)$^+$-specific glutaric dialdehyde dehydrogenase, [b] NAD(P)$^+$-specific glutaric semialdehyde dehydrogenase, [c] succinic acid semialdehyde dehydrogenase.

Preparation of cell-free extracts. *R. opacus* was grown in 2 l Erlenmeyer flasks with 500 ml MM containing 2 g/l fructose. Pyridine was then successively added up to a concentration of 350 mg/ml. The cells (5 l) were harvested after 48 h by centrifugation, washed and resuspended in 50 mM Tris-HCl (pH 8,0). Cell-free extract of the strain was prepared by passage three times through a French cell press (110 x 10^6 Pa, Fa. Amicon, Silver Spring, Md., USA). Protein concentration was estimated according to Bradford (1976).

Enzyme assay. Glutaric dialdehyde dehydrogenase (EC 1.2.1.x) was measured according to Watson and Cain (1975).

Purification of NAD(P)$^+$-specific glutaric dialdehyde dehydrogenase.
Step 1. Preparation of a crude cell-free extract.
Step 2. DEAE-Sepharose column chromatography. The cell-free extract was loaded onto a DEAE-Sepharose CL-6B column (2.6 x 20 cm) that was equilibrated and washed with 20 mM Tris/HCl buffer, pH 8.3, at 4°C. The bound proteins were eluted with a linear gradient of 0 M to 1 M NaCl in 20 mM Tris/HCl buffer (pH 8.3), and

fractions of 5 ml were collected. Fractions containing glutaric dialdehyde dehydrogenase activity were combined.

Step 3. $(NH_4)_2SO_4$ fractionation. To the eluate, solid ammonium sulphate was added to 55 % saturation. The enzyme precipitated between 40 and 55% saturation.

Step 4. Hydrophobic interaction chromatography. The enzyme from the previous step was applied to a Fast-Performance Liquid Chromatography (FPLC) Phenyl-Superose HR5/5 column (Pharmacia), equilibrated with 1 M $(NH_4)_2SO_4$ in 20 mM Tris/HCl buffer (pH 8.3). After washing the column with the same buffer, elution occurred with a linear gradient of 1 M to 0 M $(NH_4)_2SO_4$ in 20 mM Tris/HCl buffer (pH 8.3).

Step 5. Gel-Filtration chromatography. To determine the native molecular weight of the enzyme, a 20 µl sample of the Phenyl-Superose pool was applied to a HPLC Bio-Sil SEC 250 column (300 x 7.8 mm, Fa. BIO-RAD) equilibrated with 50 mM sodium phosphate buffer (pH6.8), 1 mM EDTA, 5 mM $MgCl_2$, 200 mM NaCl. Proteins were eluted in the same buffer, at a flow rate of 1 ml/min. Fractions of 0.5 ml/min were collected and analysed for enzyme activity.

Synthesis of oligonucleotides. Oligonucleotides were deduced from the N-terminal amino acid sequence and manufactured by the company MWG-Biotech GmbH.

Isolation and analysis of DNA. Total genomic DNA, isolation of plasmid DNA, use of restriction endonucleases and of ligase were done by standard procedures (Sambrook et al., 1989) or the instructions of the manufacturers (Fa. Macherey-Nagel, Germany).

Hybridisation experiments. DNA restriction fragments were separated electrophoretically in 0.8% agarose gels. Denatured DNA was transferred from agarose gels, from cell colonies or from plaques to positively charged nylon membranes (pore size, 0.45 µm) and hybridised with the DIG-dd UTP marked oligonucleotide probe (Fa. Boehringer Mannheim, Germany).

Cloning of genes. Genomic DNA of *R. opacus* was partially digested with *Bam*HI, ligated to *Bam*HI-restricted pHC79 DNA and packaged with □ coat proteins by using an *in-vitro* packaging kit (Fa. Stratagene Europe, Netherlands). Phage particles were transfected into *E. coli* S17-1 as described by Hohn and Collins (1980).

RESULTS AND DISCUSSION

Enzymatic investigations in *R. opacus* showed (Brinkmann and Babel 1997) that with growth on pyridine an NAD(P)$^+$-specific glutaric dialdehyde dehydrogenase was induced, which is considered as a key enzyme in the assimilation of pyridine. Therefore, it can be concluded, that pyridine is metabolised over the same metabolic pathway as suggested for the utilization of pyridine by *Nocardia* Z1 (Watson and Cain, 1975; Fig. 1).

At first, we started with the purification of the NAD(P)$^+$-specific glutaric dialdehyde dehydrogenase. This enzyme was purified by DEAE-Sepharose column chromatography on CL-6B and Fast-Performance Liquid Chromatography (FPLC) at Phenyl-Superose HR5/5 column (Fa. Pharmacia LKB Biotechnologie GmbH, Freiburg, Germany) (Table 1). The enzyme preparation was not homogeneous (Fig. 2). The molecular weight of the partially purified enzyme determined by gel filtration is approximately 200 kDa. The N-terminal amino acid sequence was determined by Edman degradation.

Two separate oligonucleotides (CGF01 and CGF01.1) were deduced from this sequence and labelled with DIG-dd UTP (Fa. Boehringer Mannheim, Germany). Hybridizing with *Bam*HI digested genomic DNA of *R. opacus* and the DIG-dd UTP marked oligonucleotide probe CGF01 gave one significant hybridization signal with an approximately 3.3-kbp *Bam*HI-DNA fragment (Fig.3).

TABLE 1. Purification of NAD(P)$^+$-specific glutaric dialdehyde dehydrogenase from *R. opacus*.

Fraction	Total volume (ml)	Total protein (mg)	Total activity (nmol x min^{-1})	Specific activity (nmol x min^{-1} mg^{-1} protein)	Purification (- fold)	Yield (%)
Cell-free extract	60	1254	178	142	1.0	100
DEAE-sepharose	30	183	48	264	1.9	27
Phenyl-superose	24	2	16	1198	8.4	9

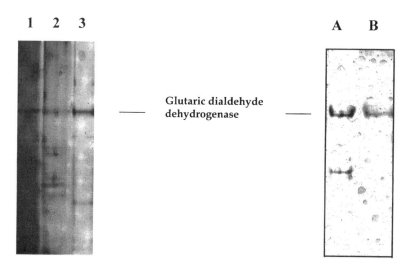

FIGURE 2a. Purification of the NAD(P)$^+$-specific glutaric dialdehyde dehydrogenase. Lane 1, crude extract; lane 2, purification of the enzyme by column chromatography on DEAE-Sepharose CL-6B; lane 3, purification by FPLC on a Phenyl-Superose HR5/5 column. 2b. PAGE of purified NAD(P)$^+$-specific glutaric dialdehyde dehydrogenase. (A) Coomassie Brilliant Blue stain. (B) Activity staining with glutaraldehyde, phenazyne methosulphate (PMS) and nitroblue tetrazolium (NBT).

The DIG-dd UTP-labelled oligonucleotide probe was used to screen a R. opacus BamHI genomic library in the cosmid pHC79 by colony hybridization. Five positive clones were detected (Fig.4). After digestion with BamHI, the hybrid cosmids pHC79::236, pHC79::268, pHC79::364, pHC79::368 and pHC79::372 contained one fragment that besides other fragments, gave a unique signal after hybridisation with the DIG labelled oligonucleotide CGF01.1: pHC79::236: 3,017 kbp BamHI-fragment, pHC79::268: 10,928 kbp BamHI-fragment, pHC79::364: 1,606 kbp BamHI-fragment, pHC79::368: 3,329 kbp BamHI-fragment, pHC79::372: 3,072 kbp BamHI-Fragment.

The fragment of the hybrid cosmid pHC79::372 was extracted from the gel and ligated into the shuttle vector pYES2 (Fig.5). The approximately 3,072 kbp BamHI-Fragment was amplified by polymerase chain reaction (PCR) and are now sequence. Final results are not present.

After amplification, i.e. increase of the copy number, physiological investigations have to show that the NAD(P)$^+$-specific glutaric dialdehyde dehydrogenase is the bottleneck.

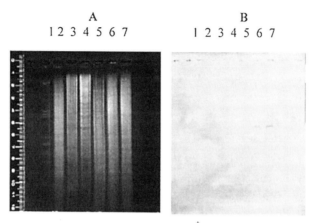

FIGURE 3. Identification of the NAD(P)$^+$-specific glutaric dialdehyde dehydrogenase gene in total genomic DNA of *R. opacus* with a DIG-dd UTP labelled oligonucleotide probe. (A) Ethidium-bromid-stained agarose gel used for blotting. (B) Blot after hybridization. *Lane* 1, molecular mass standard, *Pst*I-digested λ-DNA; *lane* 2, *Bam*HI-digested genomic DNA; *lane* 3, *Eco*RI-digested genomic DNA; *lane* 4, *Bgl*II-digested genomic DNA; *lane* 5, *Pst*I-digested genomic DNA; *lane* 6 and 7, *Bam*HI-digested genomic DNA.

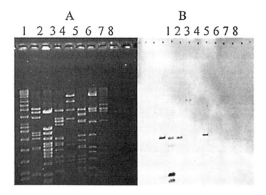

FIGURE 4. Identification of the NAD(P)$^+$-specific glutaric dialdehyde dehydrogenase gene. (A) Ethidium-bromid-stained agarose gel. (B) Blot after hybridization with the DIG-dd UTP labelled oligonucleotide probe CGF01.1. Lane 1, Raoul marker; lane 2, pHC79::372, BamHI-digested; lane 3, pHC79::417, BamHI-digested; lane 4, pHC79::236, BamHI-digested; lane 5, pHC79::268, BamHI-digested; lane 6, pHC79::364, BamHI-digested; lane 7, pHC79::368, BamHI-digested; lane 8, pHC79::476, BamHI-digested.

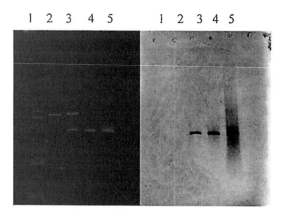

FIGURE 5. Recloning of the 3 kbp BamHI fragment, which was extracted from pHC79::372 into the shuttle vector pYES2. (Left) Ethidium-bromid-stained agarose gel. (Right) Blot after hybridization with the DIG-dd UTP labelled oligonucleotide probe CGF01.1. Lane 1, Leon marker; lane 2, pYES2, BamHI-digested; lane 3, pYES2::Ex6, BamHI-digested; lane 4, purified PCR product; lane 5, PCR of the 3 kbp BamHI fragment.

REFERENCES

Bradford M.M. 1976. A rapid and sensitive method for the quantitation of microgram quantities of protein utilizing the principle of protein dye binding. *Anal. Biochem.* 72: 248-254.

Brinkmann U., Babel W. 1996. Simultaneous utilization of pyridine and fructose by *Rhodococcus opacus* UFZ B408 without an external nitrogen source. *Appl. Microbiol. Biotechnol.* 45: 217-223.

Brinkmann U., Babel W. 1997. The role of formate for the growth of *Rhodococcus opacus* UFZ B 408 on pyridine. *Acta Biotechnol.* 17: 291-307.

Ensing J.C., Rittenberger S.C. 1963. A crystalline pigment produced from 2-hydroxypyridine by *Arthrobacter crystallopoietes* n. sp. *Arch. Mikrobiol.* 47: 137-153.

Fetzner S. 1998. Bacterial degradation of pyridine, indole, quinoline, and their derivatives under different redox conditions. *Appl. Microbiol. Biotechnol.* 49: 237-250.

Hanahan D. 1983. Studies on transformation of *Escherichia coli* with plasmids. *J. Mol. Biol.* 166: 557-580.

Hohn B., Collins J. 1980. A small cosmid for efficient cloning of large DNA fragments. *Gene* 11: 291-298.

Kaiser J.-P., Feng Y., Bollag J.-M. 1996. Microbial metabolism of pyridine, quinoline, acridine, and their derivatives under aerobic and anaerobic conditions. *Microbiol. Rev.* 60:483-498.

Korosteleva L.A., Kost A.N., Vorobéva L.I., Modyanova L.V, Terentév P.B., Kulikov N.S. 1981 Microbiological degradation of pyridine and 3-methylpyridine. *Appl. Biochem. Microbiol.* 17:276-341.

Sambrook J., Fritsch E. F., Maniatis T. 1989. *Molecular cloning: a laboratory manual*, 2nd ed. Cold Spring Harbor Laboratory, Cold Spring Harbor, N.Y.

Shukla O.P. 1973. Microbial decomposition of pyridine. *Indian J. Exp. Biol.* 11: 463-464.

Shukla O.P., Kaul S.M. 1974. A constitutive pyridine degrading system in *Corynebacterium* sp. *Indian J. Biochem. Biophys.* 11: 201-207.

Shukla O.P., Kaul S.M. 1975. Succinate semialdehyde, an intermediate in the degradation of pyridine by *Brevibacterium* sp. *Indian J. Biochem. Biophys.* 12: 326-330.

Shukla O.P., Kaul S.M. 1986. Microbiological transformation of pyridine N-oxide and pyridine by *Nocardia* sp. *Can. J. Microbiol.* 32: 330-341.

Sims G.K., Sommers L.E., Konopka A. 1986. Degradation of pyridine by *Micrococcus luteus* isolated from soil. *Appl. Environ. Mircrobiol.* 51: 963-968.

Simon R., Priefer U., Pühler A., 1983. A broad host range mobilization system for *in vivo* genetic engineering: transposon mutagenesis in gram negative bacteria. *Bio/Technology* 1:784-791.

Uma B., Sandhya S. 1997. Pyridine degradation and heterocyclic nitrification by *Bacillus coagulans*. *Can. J. Microbiol.* 43:595-598.

Watson G.K., Cain R.B. 1975. Microbial Metabolism of the pyridine ring. Metabolic pathways of pyridine biodegradation by soil bacteria. *Biochem. J.* 146: 157-172.

Zefirov N.S., Agapova S.R., Terentiev P.B., Bulakhova I.M., Vasyukova N.I., Modyanova L.V. 1994. Degradation of pyridine by *Arthrobacter crystallopoietes* and *Rhodococcus opacus* strains. *FEMS Microbiol. Lett.* 118: 71-74.

NATURAL ISOTOPE ANALYSIS: A PROMISING TOOL IN SOIL POLLUTION RESEARCH

Frank Volkering, Tauw bv, Deventer, The Netherlands
Hendrikus Jonker, Boris M. van Breukelen, Jacobus Groen, Vrije Universiteit Amsterdam, The Netherlands
Harro A.J. Meijer, University of Groningen, The Netherlands;
Barbara Sherwood-Lollar, University of Toronto, Canada,
Jan D. Kramers, University of Bern, Switzerland.

ABSTRACT: Isotopes are atoms from the same element with different atomic weights; almost all elements have more than one isotope. Analysis of the natural isotope composition of pollutants, groundwater, and geochemical species can yield unique information on pollutants and on soil and groundwater processes. This paper presents an overview of possible applications of natural isotope analysis in soil pollution research, as well as the first results of three demonstration projects in which isotope analysis has been applied: a natural attenuation assessment for a BTEX-pollution, a geochemical characterization of a landfill leachate plume, and characterization of a zinc pollution.

INTRODUCTION

Remediation strategies such as natural attenuation require detailed knowledge of soil and groundwater processes. In many cases, the information obtained via traditional characterization techniques is insufficient. Natural isotope analysis is a powerful tool providing unique information on many processes occurring in soil and groundwater. Moreover, it can be a very useful method to characterize pollutants.

Isotopes are atoms from the same element with a different number of neutrons in the nucleus, resulting in a different atomic weight. *Radioactive isotopes*, such as ^{14}C, have an unstable nucleus, causing radioactive decay with a known half-life. This makes radioactive isotopes good indicators for the age of materials, such as pollutants and groundwater. *Stable isotopes*, such as ^{13}C and the common isotope ^{12}C, are not subject to decay. Most elements on earth consist of two or more stable isotopes, behaving slightly different in various physical, chemical, and biological processes. Stable isotopes can therefore give information on the origin and fate of materials.

For analytical reasons, stable isotope concentrations are expressed using the δ-notation, relating the isotope ratio of a sample to that of a standard reference material. For ^{13}C, the standard material is the Vienna PeeDee Belemniet (VPDB), a marine carbonate. The δ for ^{13}C is defined as:

$$\delta^{13}C = \left(\frac{(^{13}C/^{12}C)_{sample}}{(^{13}C/^{12}C)_{VPDB}} - 1 \right) \times 1000 \quad (‰\ VPDB)$$

For stable isotopes of hydrogen, nitrogen, oxygen, sulfur and chlorine similar δ-notations are defined, using other standard reference materials. The change in (stable) isotope composition of molecules due to differences in reaction rates is called isotope fractionation and can often be described with the Rayleigh equation: $R = R_0 \cdot f^{(\alpha-1)}$; in which R is the isotope ratio, R_0 is the initial isotope ratio, f is the fraction residual substrate, and α is the fractionation factor. Figure 1 gives a theoretical example of the changes in isotopic composition during a fractionating reaction.

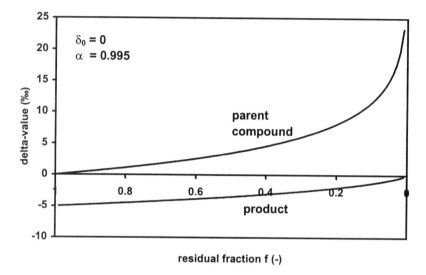

FIGURE 1. Theoretical change in stable isotope composition of substrate (parent compound) and product during a fractionating reaction.

SCOPE OF THE PROJECT

Although natural isotope analysis is a potentially useful and powerful tool, its application in soil pollution research has been limited. To promote and demonstrate the use of isotope analysis, a research project was started in May 2000. The project includes the production of a state-of-the-art on the application of isotope analysis in soil and groundwater pollution research, as well as the demonstration of different promising types of isotope analysis applications at several polluted sites.

STATE-OF-THE-ART

Isotope analysis is a rapidly developing discipline and new techniques, such as continuous flow analysis, laser spectrometry detection, and multicollector mass detection, open up new application areas and enhance the applicability existing methods (lower costs and detection limits, higher precision). Based on a literature survey, interviews with several experts, and expertise available in the project consortium, a state-of-the-art on the application of isotope analysis in soil and groundwater pollution has been compiled. An English version of this document will be available shortly (Jonker and Volkering, 2001). In the state-of the-art, four different application areas of isotope analysis are distinguished, a short description of which is given below.

Characterization of groundwater currents. Estimating the age of groundwater samples via analysis of radioactive isotopes (^{3}H, ^{3}H/^{3}He, ^{85}Kr, ^{14}C in carbonate) in groundwater is a valuable tool for calibrating and/or validating groundwater models. Stable isotopes (^{2}H, ^{18}O, ^{13}C in carbonate) may provide essential information on the origin of groundwater and thereby of local and regional groundwater flow patterns. For landfill leachates, ^{2}H and ^{3}H can be used as conservative tracers. These rather well known applications of isotope analysis are described in various textbooks (e.g. Clark & Fritz 1997).

Geochemical characterization. Many geochemical processes, such as oxidation, reduction, precipitation, and evaporation, affect the isotope composition of the compounds involved. The change in isotope composition during fractionating processes and the variation in isotopic signature of most of the lighter elements in natural materials makes stable isotope analysis suited to quantify a number of (bio)geochemical processes, such as:
- biological reactions involved in the cycles of carbon, nitrogen, and sulfur;
- physicochemical reactions such as degassing and precipitation.
Many of the processes in the carbon cycle are highly fractionating and can be characterized by measuring the ^{13}C-signature of organic carbon, carbonate species, and/or methane. In a similar way, the fractionation of ^{15}N and ^{34}S during nitrate reduction and sulfate reduction can be used to quantify these redox processes.

Pollutant characterization. For many pollutants, the isotopic signature can be used as specific "fingerprint". For organic compounds, combined ^{13}C/^{14}C-analysis can be used to distinguish between natural and anthropogenic "pollutants" and in specific cases stable isotopes can be used to relate groundwater pollutants to a source. Stable isotope analysis of nitrogen is a method to determine the origin of potential groundwater pollutants such as nitrate and ammonium. Pollutions with some specific heavy metals, such as lead, can also be characterized by their isotopic composition.

Verification of biodegradation. Stable isotopes can be used as evidence for biodegradation via two methods. Using isotopes as a fingerprint, degradation products such as CO_2 and CH_4 may be related to the original material. This method has successfully been applied to demonstrate the degradation of petroleum hydrocarbons in several field studies. The second method is based on the fractionating effect of degradation processes. This fractionation causes the residual pollution to become enriched in the heavy isotope during the course of process (parent compound in Figure 1). Compoundspecific stable isotope analysis can be applied to reveal these small but significant changes in isotopic composition.

Promising applications. Based on the information sources mentioned above, a list of promising applications for soil pollution research was made, including:
- ✓ verification of biodegradation of aromatic hydrocarbons and chlorinated aliphatic hydrocarbons via compoundspecific stable isotope analysis (^2H, ^{13}C, ^{37}Cl)
- ✓ validation of the results of hydrological modeling estimating the mean residence time (age) of recent groundwater samples via radioactive isotopes ($^3H/^3He$, ^{85}Kr)
- ✓ quantification of redox processes, such as denitrification and sulfate reduction via stable isotopes;
- ✓ characterization of pollutants, such as lead, nitrate, PAH etc., via stable isotopes
- ✓ assessment of the natural attenuation processes of cyanide via stable isotopes (^{13}C, ^{15}N)

A selection of these promising applications has been used in the three demonstration projects discussed below. Although the data presented in this paper are preliminary, the results are illustrative of the potential value of natural isotope analysis in soil pollution research..

DEMONSTRATION PROJECT 1: ISOTOPE INVESTIGATION OF A BTEX-POLLUTION AT AN INDUSTRIAL SITE

Site Description. The site investigated is a large industrial terrain in located in the Schelde estuary in The Netherlands with several groundwater pollutions and a complex hydrological setting due to the influence of tidal effects. An initial study to assess the applicability of natural attenuation as a remediation strategy for the entire location showed the groundwater to be anaerobic (sulfate reducing/ methanogenic) and provided general evidence for the natural attenuation of the aromatic and chlorinated aliphatic hydrocarbons present. The results of this initial study are presented elsewhere at this symposium (Roovers et al., 2001).

To obtain more insight in the specific degradation processes one of the pollutions has been studied in more detail. This pollution consists of two adjacent source zones near the groundwater table containing benzene and ethylbenzene and two groundwater plumes separated by a clay layer. The largest plume had a length of over 160 m. This study included the following isotope techniques:
- compoundspecific isotope analysis (CSIA) of ^2H and ^{13}C of benzene and ethylbenzene to assess biodegradation of these compounds;
- ^{13}C- and ^{14}C-analysis of methane and DOC, DIC, and calcite to verify the origin of methane;

In this paper, only the compoundspecific ^{13}C analysis will be addressed.

Results and Discussion. The highly contaminated area of the groundwater pollution was characterized by clearly elevated methane concentrations and by extremely low SO_4/Cl ratio's (characteristic for sulfate reduction), indicating degradation of ethylbenzene and/or benzene under methanogenic and/or sulfate reductive conditions to occur. Since ethylbenzene is readily degradable under anaerobic conditions and the length of the benzene-plume (over 150 m) was considerably more extended than that of the ethylbenzene-plume, ethylbenzene was concluded to be degraded. Degradation of benzene could not be ascertained using standard natural attenuation assessment techniques.

The first part of the isotope study consisted of a limited number of compoundspecific ^{13}C-analyses on groundwater samples containing benzene and ethylbenzene. The results provided supporting evidence for the presence of two separate source zones with two distinct groundwater pollutions. Benzene at the front of the largest plume (160 m from the source zone) was enriched in ^{13}C compared to benzene in the source zone (Table 1). This isotopic shift of 1.5‰ is higher than the error margins of the analysis and can be seen as an indication for biodegradation. For a reliable interpretation of the results, however, more data are needed. Therefore, a second series of compoundspecific stable carbon isotope analyses on a large number of groundwater samples containing benzene and ethylbenzene was conducted. This series also included compoundspecific deuterium analysis on benzene and ethylbenzene, a novel analysis technique with exciting perspectives for assessing biodegradation of organic compounds (Mancini et al., 2001). The results of these combined compoundspecific analyses will be available during the time of this presentation and are expected to provide a conclusive answer to the question of occurrence of anaerobic benzene degradation at the site.

TABLE 1. Concentration and stable carbon isotope results for benzene and ethylbenzene in selected groundwater samples

location	depth (m bgs)	benzene concentration (μg/L)	benzene ^{13}C-content (‰ VPDB)	ethylbenzene concentration (μg/L)	ethylbenzene ^{13}C-content (‰ VPDB)
source zone A	2.0-3.5	92,000	-28.6± 0.5	44,000	-28.4± 0.5
source zone B	1.5-2.5	62,500	-29.5± 0.5	35,500	-28.9± 0.5
center plume B	11.5-12.5	*54,000	-29.2± 0.5	99	n.d.
front plume B	11.5-12.5	1,560	-28.0± 0.5	4	n.d.

* concentration measurement unreliable due to interfering compounds
n.d.: concentration below detection limit for stable carbon isotope analysis

DEMONSTRATION PROJECT 2: ISOTOPE INVESTIGATION OF A LANDFILL LEACHATE PLUME

Site Description. The leachate of the Dutch Banisveld landfill site has been the subject of a detailed geochemical investigation that will be presented elsewhere at this symposium. (Van Breukelen et al., 2001b). The landfill leachate contains high concentrations of DOC and several ionic species, as well as low concentrations of aromatic micropollutants, such as benzene and naphthalene. Geophysical techniques have been applied to delineate the vertical and horizontal distribution of the leachate plume. Based on geochemical and biochemical analyses, iron reduction was concluded to be the dominating redox process in the plume. Groundwater below the plume is methanogenic, whereas in the groundwater above the plume (near the saturated zone) denitrification is the dominating redox process.

To verify the conceptual model, an isotope study was performed, including isotopic analysis of groundwater (^2H, ^{18}O), methane (^2H, ^{13}C), DIC, DOC (^{13}C), and sulfate (^{34}S) via standard methods.

Results and Discussion. This paper will only discuss the isotope results for methane. The ^2H and ^{13}C content provide information on the origin of methane. Thermogenic methane, methane produced via CO_2-reduction, and methane produced via acetate fermentation all have a characteristic isotopic composition, as is shown in Figure 2. This figure also includes the experimental data from the Banisveld study.

Methane occurring naturally in the groundwater below the leachate plume (concentration 1.5 mg/L) has a isotopic signature characteristic for methane produced via CO_2-reduction (Hackley et al. 1996). This is on, as CO_2 is assumed to be the dominating methanogenic process in groundwater (Clark & Fritz 1997). Methane inside the plume, present in concentrations of 14-23 mg/L, has a isotopic composition typical for acetate fermentation and landfill gas. This methane in the plume was concluded to originate from the anaerobic processes occurring inside the landfill. Methane measured in the denitrifying groundwater above the plume was lower in concentration (4.9 mg/L) and slightly enriched in ^{13}C, indicating oxidation of the landfill-derived methane (a fractionating process).

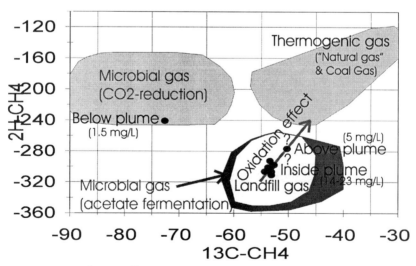

FIGURE 2. ^2H and ^{13}C of methane from different sources, markers (•) represent data from the Banisveld study

DEMONSTRATION PROJECT 3: CHARACTERIZATION OF A ZINC POLLUTION VIA STABLE ISOTOPES

Site description. In the Dutch town of Ede, a small canal is used for the discharge of zinc-contaminated water by a viscose production facility and by the local wastewater treatment plant. The sediment of the canal is regularly dredged, and in the past zinc-contaminated dredged material has been deposited on the benches of the canal and possibly at some other unknown locations in Ede. To establish the possibility of zinc characterization via stable isotopes, stable zinc analyses (Zn-64, Zn-66, Zn-67, Zn-68) were performed by the University of Bern (Switzerland) based on the method described by Marechal et al. (2000). This method has been used for analyzing samples from the two different effluents involved and for groundwater samples from monitoring wells located near the canal, upstream of the wastewater treatment plant.

Results and Discussion. The results of first set of stable zinc analyses show a significant difference between the zinc from the two sources in Ede. Figure 3 gives a graphical representation of the preliminary results.

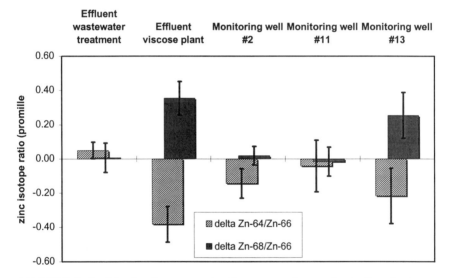

FIGURE 3. Stable zinc isotope ratios from several samples in ‰ relative to the J&M zinc standard

Zinc from the wastewater treatment plant originates mainly from zinc-plated surfaces and is rather similar to the zinc-standard used (J&M standard). The zinc from the viscose production facility originates from zinc sulfate and is relatively depleted in Zinc-64 and enriched in Zinc-68. The isotopic signature of the zinc from monitoring well #13 shows a good agreement with the zinc from the viscose production facility; zinc from wells #2 and #11 is rather similar to zinc originating from the wastewater treatment and is likely to originate from zinc-plated surfaces. Additional zinc isotope analysis of samples from other monitoring wells and from the sediment of the canal is expected to provide more information on the origin of zinc in the monitoring wells. To our knowledge, this is the first time natural isotope analysis has been applied as a characterization method for zinc in the environment.

ACKNOWLEDGEMENTS

This project partly was financed by the Dutch Soil Research Program (SKB), by Dow Chemicals, and by the Dutch provinces of Drenthe, Gelderland, and Noord-Brabant

REFERENCES

Clark I.D., and P. Fritz. 1997. *Environmental Isotopes in Hydrogeology*, Lewis Publishers, New York.

Hackley K.C., C.L. Liu, and D.D. Coleman. 1996. "Environmental Isotope Characteristics of Landfill Leachates and Gases". *Ground Water* 34(5):827-836

Jonker H, and F. Volkering. 2001. *State-of-the art on the Application of Isotope Analysis in Soil Pollution Research*. SKB-report to appear in 2001.

Mancini S.A., A.C. Stelfox, J.A.M. Ward, J. Ahad, G. Lacrampe-Couloume, B. Sleep, E. Edwards, and B. Sherwood Lollar. 2001. "Stable Carbon and Hydrogen Isotope Fractionation During Anaerobic Biodegradation of Aromatic Hydrocarbons". *Material scheduled to be presented at the Sixth International In Situ and On-Site Bioremediation Symposium (San Diego, June 4-7, 2001)*

Marechal, C.N., P. Telouk, and F. Albarede. 1999. "Precise Analysis of Copper and Zinc Isotopic Compositions by Plasma-source Mass Spectrometry". *Chemical Geology* 156: 251-273

Roovers C.P.A.C, S.J.T. Eskes, and P. van Riet. 2001. Optimized Application of Natural Attenuation: The Buffer Zone Remediation Concept. *Material scheduled to be presented at the Sixth International In Situ and On-Site Bioremediation Symposium (San Diego, June 4-7, 2001)*

Van Breukelen B.M., W.F.M. Roling, J. Griffioen, and H.W. van Verseveld. 2001. "Biogeochemistry of a Landfill Leachate Plume, Consequences for Natural Attenuation". *Material scheduled to be presented at the Sixth International In Situ and On-Site Bioremediation Symposium (San Diego, June 4-7, 2001)*

CHARACTERISATION OF MICROBIAL *IN SITU* DEGRADATION OF AROMATIC HYDROCARBONS

Hans H. Richnow, Matthias Gehre, Matthias Kästner, (UFZ- Centre for Environmental Research, Leipzig, Germany)
Barbara Morasch, (University of Konstanz, Konstanz, Germany)
Rainer U. Meckenstock, (Eberhard-Karls University of Tübingen, Tübingen, Germany)

ABSTRACT. Carbon isotope fractionation was studied in a soil percolation column. The soil microbial community was able to degrade toluene generating a stable concentration gradient within the column. The carbon isotope composition of the residual non-degraded toluene fraction showed a significant increase in the $^{13}C/^{12}C$ ratio. The extent of biodegradation in this soil column was calculated using the measured isotope ratios (R_t) and an isotope fractionation factor (αC = 1.0017) obtained from a sulfate-reducing batch culture. The calculated residual substrate concentrations (C_t) were in good accordance with to the measured toluene concentrations in the column.

The isotopic composition of BTEX compounds was studied in contaminated anoxic aquifers. Along the contamination plume, decrease of contaminant concentrations was associated with a distinct increase of $\delta^{13}C$ values, suggesting *in situ* biodegradation. The contribution of microbial degradation to the total contaminant removal was calculated based on laboratory-derived carbon isotope fractionation factors using the Rayleigh equation. The results show that microbial degradation was the most important contaminant removal process in the aquifer.

INTRODUCTION

Assessment of the *in situ* biodegradation of pollutants is needed for a prediction of the fate of contaminants in aquifers. Decreasing concentrations of a pollutant downstream of the source of contamination may may be due to dilution, volatilization, adsorption to soil, and biodegradation by the indigenous microflora. To characterize the *in situ* microbial degradation of aromatic hydrocarbons, we used isotope fractionation of contaminants during biodegradation (Richnow and Meckenstock, 1999).

Carbon isotope fractionation during toluene degradation by aerobic and anaerobic bacteria was observed in previous studies (Meckenstock et al, 1999; Ahad, et al. 2000). With decreasing concentrations, the carbon isotope ratio in the residual non-degraded toluene fraction increased. The Rayleigh equation can be applied to calculate the carbon isotope fractionation factors (αC) (Hoefs, 1997) and was determined for toluene degradation by *P. putida* strain mt2 (1.0027), *T. aromatica* (1.0017), *G. metallireducens* (1.0018), and by the sulfate-reducing strain TRM1 (1.0017). Although these organisms used different electron acceptors for toluene oxidation, the kinetic isotope fractionation factors αC were within the same order of magnitude (Meckenstock et al., 1999). Non-biological isotope effects, which may be caused in a multiphase system by transitions between the liquid and the gas phase or isotope effects due to sorption to soil, were not observed (Meckenstock et al., 1999; Slater at al., 2000). Recent studies on the biodegradation of chlorinated hydrocarbons with laboratory cultures revealed carbon isotope fractionation of these compounds and field site evidence showed that carbon isotope fractionation also takes place in aquifers. The fractionation can be used as an indicator for biodegradation (Dempster et al., 1997; Sherwood Lollar, et al., 1999; Bloom et al., 2000; Hunkeler et al., 1999).

Objective. The aim of this study was to demonstrate that the Rayleigh equation could be used to calculate the percentage of biodegradation of the residual substrate fraction. Moreover, under specific conditions the amount of the biodegraded fraction could be quantified. Compound-specific isotopic fractionation factors (αC) of BTEX compounds were determined in microbial degradation experiments in the laboratory. These factors were used to calculate *in-situ* biodegradation based on variation of the isotopic signatures of contaminants within an aquifer.

MATERIALS AND METHODS

Field study 1. The contaminated aquifer was located in an industrial area near the city of Hamburg, Germany. The contamination dated back more than 30 years and was a result of leaking storage tanks containing an aromatic oil rich in naphthalene. The site was equipped with a number of monitoring wells along the groundwater flow direction. The groundwater flow generated a contamination plume more than 800 m long. The contaminated zone is oxygen free suggesting that anoxic conditions prevailed (Richnow et al. 2001b)

Field study 2. Vejen Landfill, Denmark, is an old landfill of mixed municipal and industrial waste leaking into a naturally oxic shallow sandy aquifer. The leachate has entered the groundwater since approximately 1973 and has formed a plume stretching 370 m down gradient from the landfill (Lyngkilde and Christensen, 1992a,b). Most of the organic contaminants were degraded under strongly reducing conditions (methanogenic, sulfate-reducing and iron-reducing) within the first 50 m of the plume. A transect in this part was investigated in detail (Richnow et al. 2001a).

Soil percolation columns. Soil column studies were performed in glass columns filled with contaminated aquifer material and equipped with 5 sampling ports at 2.5 cm, 7 cm, 16 cm, 26 cm and 36 cm distance from the inlet. The bottom-to-top water flow was adjusted with peristaltic pumps to one bed volume per day and operated in the absence of light at constant temperature (16°C) for several months with a mineral medium containing 32 mg toluene L^{-1} and sulfate as electron acceptor (Meckenstock et al., 1999).

GC analyses. Groundwater samples were extracted with *n*-pentane and solvent extracts were quantified by gas chromatography. A GC-6000 Vega 2 instrument

(CARLO ERBA INSTRUMENTS, Milan, Italy) equipped with a fused silica capillary column (DB-5, 60 m x 0.25 mm I.D. x 0.25 µm film, J&W SCIENTIFIC, USA) and a flame ionization detector was used (Richnow et al., 2001b).

GC-C-IRMS analysis. The carbon isotope composition of aromatic hydrocarbons was measured with a GC-C-IRMS (gas chromatography/combustion/isotope-ratio-monitoring mass spectrometry) system (FINNIGAN MAT, Bremen, Germany). The GC-C-IRMS system consisted of a GC unit connected to a FINNIGAN MAT combustion device with a water removal assembly coupled to a FINNIGAN MAT 252 mass spectrometer. All samples were measured in at least 5 replicates. Analytical details are given elsewhere (Richnow et al., 2001b).

Calculations. All carbon isotope ratios are given in the delta notation as $\delta^{13}C$ values [‰] which are related to PDB (Pee Dee Belemnite) as standard. R is the isotope ratio $^{13}C/^{12}C$ calculated in Equation (1).

$$\delta_t\ ^{13}C\ ‰ = (\frac{^{13}C/^{12}C\ \text{sample}}{^{13}C/^{12}C\ \text{standard}} - 1) \times 1000 = (R_t/R_{Std} - 1) \times 1000 \quad (1)$$

$$R_t/R_0 = (\delta_t + 1000)/(\delta_0 + 1000) \quad (2)$$

$$R_t/R_0 = f_t^{(1/\alpha C - 1)} \quad (3)$$

$$\ln(R_t/R_0) = (1/\alpha C - 1) \times \ln f_t \quad (4)$$

$$f_t = (R_t/R_0)^{(1/(1/\alpha C - 1))} \quad (5)$$

$$C_t = C_0 \times f_t \quad (6)$$

$$Bf_t\ [\%] = (1 - f_t) \times 100 \quad (7)$$

Kinetic isotope fractionation factors αC were calculated using Equation 4, which is derived from the Rayleigh equation for a closed system (Equation 3) (Hoefs, 1997). δ_t is the carbon isotope composition at time t, δ_0 is the initial carbon isotope composition of the substrate, and f_t is the fraction of substrate remaining in the sample at time t ($f_t = C_t/C_0$). The slope of the linear regression curve gives the kinetic isotope fractionation factor αC as $(1/\alpha C - 1)$ (Equation 4) when $\ln(R_t/R_0)$ is plotted over $\ln(C_t/C_0)$ for the time intervals t. The remaining substrate fraction $f_t = C_t/C_0$ can be calculated using the isotope fractionation factor (αC) obtained in laboratory experiments (Equation 5 and 6). The percentage of biodegradation of the residual substrate fraction (Bf_t) is calculated with Equation 7.

RESULTS

Laboratory studies. Non-sterile, soil percolation column experiments were conducted in order to study the sulfate-dependent anaerobic degradation of toluene under conditions similar to those prevailing in aquifers. A microbial community was established which was able to degrade toluene to a concentration below 0.05 mg L^{-1}, generating a stable concentration gradient of toluene along a flow path of 26 cm in the percolation column (Figure 1A).

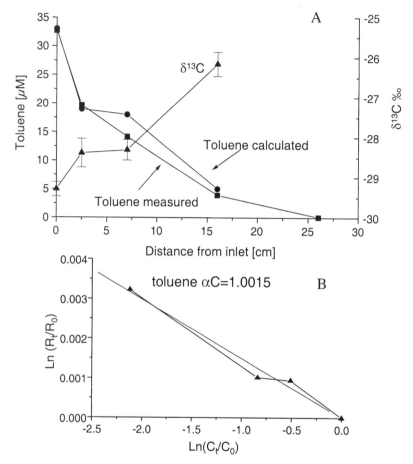

FIGURE 1. (A) Anaerobic toluene degradation in soil percolation column showing concentrations (squares) and $^{13}C/^{12}C$ isotope compositions (triangles) of toluene (solid line). The residual concentrations of toluene (dots) were calculated using the measured isotope ratios and a carbon isotope fractionation factor of 1.0017.
(B) The slope of the regression of toluene concentrations over the respective $^{13}C/^{12}C$ isotope ratios according to Equation 4 gives the isotope fractionation factor for the toluene degradation in the column (modified after Richnow et al., 2001b).

Analysis of the carbon isotope ratio of the residual toluene revealed a significant increase in $\delta^{13}C$ with decreasing concentrations. To obtain the isotope fractionation factor αC, the logarithm of concentration ratios (C_t/C_o) was plotted versus the logarithm of isotope ratios (R_t/R_o) in accordance to Equation (4) (Figure 1B). The linear regression of the curve give the $^{13}C/^{12}C$ isotope fractionation factor for toluene degradation in the column. The isotope fractionation factor ($\alpha C = 1.0017$) of toluene degradation from batch cultures with sulfate as electron acceptor was applied to calculate the extent of biodegradation in the soil percolation column experiment using Equations 5 and 6. The enrichment of ^{13}C in the residual toluene fractions (R_t) was used to calculate the theoretical residual concentrations (C_t) and the calculated concentrations and match the measured concentrations of toluene in the soil percolation column quite well (Figure 1A).

The calculation is based on a mathematical description of isotope fractionation by the Rayleigh equation for closed systems (Rayleigh, 1896). Closed system conditions are certainly realized in biodegradation experiments with pure cultures in closed systems where the only loss of substrate is caused by microbial degradation. Abiotic factors such as sorption and dilution influence the concentration contaminants did not affect the isotope composition to a significant extent (Meckenstock et al., 1999; Slater et al., 2000; and unpublished data). Although a soil percolation column is not a closed system, under constant input condition, the removal of contaminants is only a result of biodegradation. Dilution and adsorption can be neglected and the concentration and isotopic composition of the substrate at the inlet of the column would remain constant along the water flow path when no biodegradation occurs. In microbially active columns, the concentration and isotope composition vary over the distance from the inlet instead of time as within batch cultures. Time may be substituted by distance and the Rayleigh equation for closed systems can be applied to calculate the isotope fractionation factor α. Vice versa, the isotope fractionation factor can be applied to calculate the degraded fraction of the substrate by the shift in the isotopic composition. In the soil percolation column experiment, the calculated amounts of the residual toluene fractions matched the measured concentrations quite well when a carbon isotope fractionation factor ($\alpha C = 1.0017$) of a toluene degrading, sulfate-reducing batch culture was used. This showed that the Rayleigh equation could be used to quantify biodegradation in the soil column where dilution and sorption could be neglected as relevant sinks for contaminants.

Field study 1. The carbon isotope ratios and concentrations of aromatic contaminants such as BTEX were analyzed along the 800 m profile downstream of the contamination source. Toluene showed a significant concentration gradient in the plume decreasing from 160 to 1.9 $\mu g\ L^{-1}$ along the profile and dropped below the detection limit of 0.5 $\mu g\ L^{-1}$ in observation well F (Figure 2). With decreasing concentrations, the $^{13}C/^{12}C$ isotope ratio in the residual toluene fraction was enriched up to 7.2 ± 1.0‰ between wells A and E (Figure 2). The concentrations and isotopic compositions of contaminants in the field were used to examine *in situ* biodegradation at the test site by applying the Rayleigh equation and fractionation factors determined in laboratory experiments. To calculate the extent to which the toluene fraction was degraded by microorganisms, the carbon isotope fractionation factor for anaerobic toluene degradation in the soil column ($\alpha C = 1.0015$) was used. The toluene isotopic composition at well A ($\delta^{13}C = -23.0$ ‰)

showed the lowest measured $\delta^{13}C$ value compared to the other monitoring wells. Thus, this isotope value was used as the initial isotopic composition of a theoretical source (R_0) to calculate the extent of the biodegradation (Bf_t) (Figure 2). Bf_t (percentage of the biodegradation) is an index characterizing the extent of biodegradation along the water flow path but is not quantitative in terms of the actual contaminant concentrations. However, applying the isotope value $\delta^{13}C$ at well C, the remaining toluene fraction had been biodegraded by about 54% (Bf_t). Further downstream of this well, the calculation indicates that the percentage of biodegradation (Bf_t) had increased to more than 99% at well E (Figure 2).

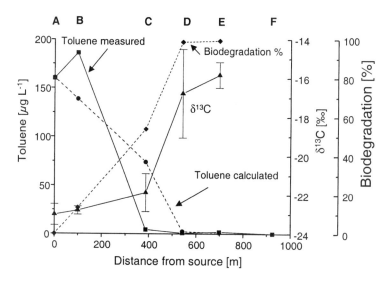

FIGURE 2 Concentrations (squares) and $^{13}C/^{12}C$ isotope ratios (triangles) of toluene along the monitoring profile (A – E) of a contaminated aquifer (Hamburg, Germany). Circles indicate the percentage of biodegradation (Bf_t). Diamonds indicate the extent to which the residual toluene fraction has been degraded (modified after Richnow et al., 2001b).

The extent of biodegradation of the residual substrate fraction can be quantified, assuming that (I) all monitoring wells were located directly downstream of the source of contamination, (II) the contaminants were bioavailable, and (III) dilution and sorption could be neglected. Based on these assumptions, the isotopic compositions and concentrations monitored in well A could be employed as R_0 and C_0 to calculate the absolute amount of biodegraded toluene in wells B–E (equation 5 and 6). The residual toluene concentrations calculated showed a slow decrease for wells B and C, only roughly matching the residual concentrations measured (Figure 2). However, further downstream the groundwater flow path the calculated concentrations decreased strongly, reflecting the measured toluene concentrations in the wells.

Field study 2. In a second field study the concentration and isotopic composition of BTEX was analyzed in an anaerobic part of a contamination plume formed by the leachate of an old landfill (Vejen, Denmark). The concentration of ethylbenzene in the aquifer varied down gradient between 123 and 2 µg L^{-1} The isotopic composition of ethylbenzene showed an enrichment in δ^{13}C from –26.3 ± 0.63 up to –17.5 ± 0.86‰ PDB (Figure 3). Strikingly, the highest concentration (123 µg L^{-1}) and most depleted isotopic composition (–26.3 ± 0.63 PDB) were found down gradient, suggesting that this material showed the lowest degree of biodegradation on this test site, and is a relic of a former plume (Richnow et al., 2001a). In other parts of the plume, more intensive biodegradation has lead to an enrichment of ^{13}C in the residual ethylbenzene fraction. The concentration and isotopic composition of ethylbenzene compiled with the Rayleigh plot (equation 4) gave an isotope fractionation factor of αC = 1.0021 with a reasonable correlation (R^2 = 0.6406) (Richnow et al., 2001a), suggesting that concentration and isotopic composition of ethylbenzene were significantly controlled by biodegradation. To calculate the percentage of biodegradation (Bf$_t$), an isotope fractionation factor (αC = 1.0039) from a denitrifying ethylbenzene-degrading culture (data not shown) was applied. The most negative δ^{13}C-value was used as R$_0$. The curve of the percentage of biodegradation shows the extent of biodegradation in the various zones of the cross section analyzed. Moreover, the theoretical concentration of the residual fraction was quantified using 123 µg L^{-1} as C$_0$. The curve of the calculated concentrations was always higher than the measured concentrations indicating that other processes also influenced the distribution of contaminants on the site (Figure 3). However, the calculation suggests that major parts of the ethylbenzene were removed by *in situ* biodegradation.

DISCUSSION AND CONCLUSION

A concept is presented to estimate the amount of the biodegraded fraction of contaminants in an aquifer using stable isotope fractionation. Because an aquifer is not a closed system, the use of the Rayleigh approach to calculate the extent of biodegradation may run up against some limitations, which were considered when interpreting the results.

In addition to biodegradation, the concentration of the contaminants could be deminished along the groundwater flow path by mixing with pristine groundwater, adsorption to the aquifer matrix, and evaporation. Evaporation was considered to be of minor influence as a significant loss of contaminants to the vadose zone is restricted because turbulent mixing is unlikely in aquifers and diffusion is a very slow process. Moreover, in deeper aquifers the geological situation may prevent a significant loss of contaminants due to volitalization.

Dilution does not alter the carbon isotopic composition of substrates, and adsorption had no effect on the carbon isotope ratios of aromatic hydrocarbons in soil percolation columns (Meckenstock et al., 1999). Therefore, the extent of biodegradation of the residual substrate fraction can be calculated using the difference in the isotopic compositions of the contaminants downgradient along the groundwater flow path. The calculation gives the percentage of biodegradation (Bf$_t$) necessary to change the isotopic composition of the residual fraction.

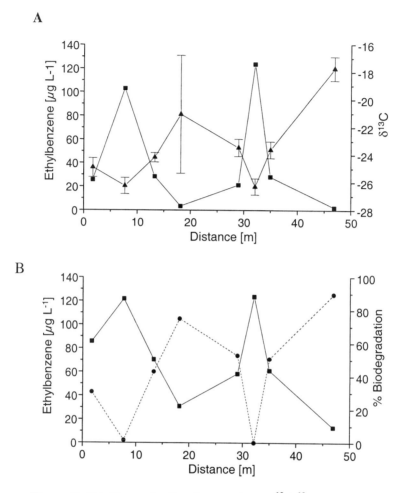

Figure 3. (A) Concentration (squares) and $^{13}C/^{12}C$ isotope ratios (triangles) of ethylbenzene along the monitoring profile of a contaminated aquifer (Vejen, Dennmark). (B) Concentration of the residual ethylbenene fraction (squares) as calculated based on the carbon isotope ratios. Circles indicate the percentage of biodegradation (Bf_t). (data from Richnow et al., 2001a)

However, the index "percentage of biodegradation" can be influenced by a mixing of water flow paths with a different extents of biodegradation. For example, in a worst case situation, contaminants would be completely degraded in one flow path, whereas no degradation would occur in another flow path. After both water bodies are mixed in a well, the isotope ratios of the contaminants would show the initial isotopic composition of the source, because the substrate from the flow path which does not exhibit biodegradation would have a much higher concentration and would dominate the isotope ratio of the mixed sample. As a result of this example scenario, although significant amounts of substrate could have been degraded, calculation would indicate only minor biodegradation, underestimating its true extent. However, the calculation will not overestimate the extent of biodegradation regardless of mixing processes in the aquifer or sorption to the aquifer's matrix, because these processes did not influence the isotopic composition of contaminants. To select an appropriate isotope fractionation factor for the calculation information of the biochemical degradation conditions such as temperature, redox and electron acceptors in the contaminated aquifer must be available, because the biochemical degradation pathway influence the extent of isotopic fractionation.

Adsorption and dilution restrict the use of the Rayleigh equation in quantifying the absolute amount of the biodegraded substrate fraction. Assuming that (I) the aquifer is in steady state with respect to adsorption or desorption of contaminants, (II) the total substrate concentration is bioavailable, and (III) dilution is insignificant, the initial isotope ratio (R_0) and the substrate concentration (C_0) can be used as the initial values for all wells along the groundwater flow path. In this case, the Rayleigh equation could be used to calculate the theoretical residual substrate concentration in the wells (C_t) down gradient the water flow path using the isotopic composition of the residual fraction (R_t). This may illustrate also the complications to quantify the *in situ* biodegadation when using the isotopic fractionation concept.

REFERENCES

Ahad, J. M. E., B. Sherwood Lollar, E. A. Edwards, G. F. Slater, and B. E. Sleep. 2000. "Carbon isotope fractionation during anaerobic biodegradation of toluene: Implications for intrinsic bioremediation." *Environ. Sci. Technol.* 34: 892-896.

Bloom, Y., R. Aravena, D. Hunkeler, E. Edwards, and S. K. Frape. 2000. "Carbon isotope fractionation during microbial dechlorination of trichloroethene, cis-1,2-dichloroethene, and vinyl chloride: Implications for assessment of natural attenuation." *Environ. Sci. Technol.* 34: 2768-2772.

Dempster, H., B. Sherwood Lollar, and S. Feenstra. 1997. "Tracing organic contaminants in groundwater: A new methodology using compound-specific isotopic analysis." *Environ. Sci. Technol.* 31: 3193-3197.

Hoefs, J., 1997. *Stable isotope geochemistry*. 4th edn. Springer Verlag, Berlin, Germany.

Hunkeler, D., R. Aravena, and B. J. Butler. 1999. Monitoring microbial dechlorination of tetrachloroethene (PCE) in groundwater using compound-specific stable carbon isotope ratios: microcosm and field studies. *Environ. Sci. Technol.* 33: 2733-2738.

Lyngkilde, J., and T.H. Christensen. 1992a. "Redox zones of a landfill leachate pollution plume (Vejen, Denmark)". *J. Contam. Hydrol.* 10: 273-289.

Lyngkilde, J., and T.H. Christensen. 1992b. "Fate of organic contaminants in the redox zones of a landfill leachate pollution plume (Vejen, Denmark)." *J. Contam.*

Hydrol. 10: 291-307.

Meckenstock, R.U., B. Morasch, R. Wartmann, B. Schink, E. Annweiler, W. Michaelis, and H. H. Richnow. 1999. "$^{13}C/^{12}C$ isotope fractionation of aromatic hydrocarbons during microbial degradation." *Environ. Microbiol.* 1: 409-414.

Richnow H.H., and R.U. Meckenstock. 1999. "Isotopen-geochemisches Konzept zur *in-situ-* Erfassung des biologischen Abbaus in kontaminiertem Grundwasser." *TerraTech* 1999: 38-41.

Richnow H.H., E. Annweiler, W. Michaelis, B. Schink and R.Meckenstock. 2001b. "Microbial *in situ* degradation of aromatic hydrocarbons in a contaminated aquifer monitored by carbon isotope fractionation." *J. Contam. Hydrol.* (submitted)

Richnow H.H., R.U. Meckenstock, L. Ask, A. Baun, A. Ledin, and T.H. Christensen. 2001a. "*In situ* biodegradation determined by isotope fractionation of aromatic hydrocarbons in an anaerobic landfill leachate plume (Vejen, Denmark)." *J. Contam. Hydrol.* (submitted)

Sherwood Lollar, B., G. F. Slater, J. Ahad, B. Sleep, J. Spivack, M. Brennan, and P. MacKenzie. 1999. "Contrasting carbon isotope fractionation during biodegradation of trichloroethylene and toluene: Implications for intrinsic bioremediation." *Org. Geochem.* 30: 813-820.

Slater, G.F., J.M.E. Ahad, B. Sherwood Lollar, R. Allen-King, and B. Sleep. 2000. "Carbon isotope effects resulting from equilibrium sorption of dissolved VOCs." *Anal. Chem.* 72, 5669-5672.

IN SITU VINYL CHLORIDE BIODEGRADATION REVEALED THROUGH CARBON ISOTOPE COMPOSITION

Duane Graves, Ph.D., IT Corporation, Knoxville, TN, USA; Gary R. Hecox, Kyle Kirschenmann, and Sherry Ingram, IT Corporation, Baton Rouge, LA, USA; Barbara Sherwood Lollar, Ph.D., University of Toronto, Toronto, Canada

ABSTRACT: In situ biodegradation of vinyl chloride (VC) was documented by the change in the stable carbon isotope composition of VC within a groundwater aquifer. Isotopic specificity has been documented for stable carbon isotopes during the biodegradation of chlorinated ethenes, including VC. Biodegradation involves a preferential degradation of isotopically light ^{12}C-bearing VC. The preferential reaction results in accumulation of ^{13}C in the residual VC such that the isotopic composition of the VC becomes measurably ^{13}C enriched.

This phenomenon was used to evaluate the effects of biodegradation on a large VC plume in a deep sandy aquifer at a manufacturing facility. Analysis of the stable carbon isotope composition of tetrachloroethene (PCE), trichloroethene (TCE), and VC demonstrated that VC was being biodegraded, provided an estimation of the degree of biodegradation in the source area and down gradient areas of the plume, and supported two independent biodegradation rate calculations that did not rely on VC concentration data.

Stable carbon isotope analysis provided dramatic evidence for biodegradation and allowed an estimation of the degree and rate of VC biodegradation. The occurrence of biodegradation would have been assumed based on site specific geochemical changes and the presence of ethene in the groundwater. However, the significance of VC biodegradation would have been severely underestimated using concentration data alone and biodegradation rate calculations could not have been determined with existing data.

INTRODUCTION
Natural biodegradation of chlorinated ethenes was evaluated using a combination of routine groundwater analyses and innovative stable isotope measurements. The combination of these techniques provided definitive evidence of biodegradation. Complexities of the site prevented rate calculation estimations using typical first-order decay equations. The stable carbon isotope data were particularly useful determining VC degradation rate calculations.

The site was located near an estuary on a Pleistocene Age sequence of clay and silty clays with thin, interbedded lenticular sand intervals. These intervals are present as a series of five low-yield water-bearing units (WBU) at approximate depths of 10, 20, 36 to 50, 70 to 80, and 120 feet below ground surface. In the south part of the site, the intervals at 10, 20, and 36 to 50 feet deep were absent, having been replaced by younger sediments and fill material referred to as the Unconfined WBU and the Semiconfined WBU. The regional aquifer system consisted of three primary sand units, the 200-Foot Sand, the 500-Foot Sand, and the 700-Foot Sand at depths of between 150 and 900 feet below ground surface. This investigation quantified the effect of natural attenuation processes on VC concentration and persistence in the 200-Foot Sand of the regional aquifer.

METHODS

Sampling locations were selected to represent background conditions in the groundwater and nearby estuary, conditions within and down gradient of the VC-impacted area of the 200-Foot Sand, and conditions above the 200-Foot Sand representing the groundwater and chlorinated solvents moving downward into the 200-Foot Sand (Table 1). Groundwater samples from these wells and the estuary were analyzed for chlorinated ethenes, natural attenuation monitoring parameters, and stable carbon and chlorine isotopes. The results were evaluated for evidence of natural attenuation and intrinsic biodegradation and to estimate the rates for degradation of VC.

Samples were collected after the wells had been purged of the standard three casing volumes. A flow-through cell was used for field measurements of dissolved oxygen, oxidation-reduction potential, pH, conductivity, and temperature. Field colorimetric tests were used to measure the concentrations of ferrous iron, sulfide, and alkalinity. Laboratory-based groundwater analyses consisted of volatile organic compounds (VOC) using US EPA SW-846 Method 8260, nitrate, nitrite, sulfate, methane, ethene, ethane, total organic carbon, chloride, dissolved hydrogen, stable carbon isotopic composition of the VC, and stable isotopic composition of chlorine in groundwater.

The Stable Isotope Laboratory at the University of Toronto, Toronto, Canada, analyzed samples for stable carbon isotopes using a gas-chromatograph combustion isotope-resolving mass-spectrometer (GC/C/IRMS). The Environmental Isotopes Laboratory at the University of Waterloo, Ontario, Canada, analyzed the samples for stable chloride isotopes.

RESULTS AND DISCUSSION

The field and laboratory data were evaluated using approved procedures recommended by the US EPA (US EPA, 1998). Additionally, the results from the isotope analyses were evaluated based on published findings relative to the behavior of stable isotopes during biological reactions (Aravena et al., 1998; Bloom et al., 2000; Slater et al., 2000; Sherwood Lollar et al., 1999). The results provided strong evidence of biodegradation of VC in the regional aquifer. Direct evidence indicating the occurrence of intrinsic biodegradation of chlorinated ethenes including VC was derived from an examination of the chemical composition of chlorinated ethenes in groundwater, the $\delta^{13}C$ of PCE, TCE, VC, and ethene, and the $\delta^{37}Cl$ in background and VC-impacted groundwater.

Chlorinated Ethene Composition in Groundwater. The chemical constituency of dissolved chlorinated ethenes in the groundwater revealed a marked shift in the composition of the chlorinated ethenes from the source area to down gradient wells. The source area wells contained higher concentrations of more highly chlorinated ethenes (PCE, TCE, and DCE) than the down gradient wells. In contrast, the down gradient wells were not contaminated with chlorinated ethenes or contained only VC (Table 2). Examination of the cis- to total 1,2-DCE ratios for the three wells that contained 1,2-DCE revealed that cis-1,2-DCE was the dominant isomer. The cis-1,2-DCE to total DCE ratio exceeded 90 percent in all cases. These observations provide evidence for biologically mediated reductive dechlorination of chlorinated ethenes in the groundwater.

Ethene Concentration in Groundwater. The coincidence of high concentrations of ethene and ethane at locations that also contained high concentrations of VC strongly suggest that reductive dechlorination of VC was occurring (Table 2). A potentially complicating factor in this interpretation is the widespread presence of dissolved natural gas in the 200-Foot Sand. Methane and ethane are prominent constituents of natural gas; however, ethene is not a component of natural gas (Pedersen et al., 1989). The groundwater at the site has a background methane concentration of about 35 mg/L. All groundwater samples contained a low concentration of ethane, but only wells impacted with VC contained detectable concentrations of ethene. This observation suggests that the presence of ethene was due to biological formation through reductive dechlorination of VC rather than as a trace compound in natural gas.

The monitoring well with the highest VC concentrations also had the highest ethene (6.5 mg/L) and methane concentrations (116 mg/L), but the ethane concentration (0.012 mg/L) was no higher than that observed in the other wells (average site-wide ethane concentration, 0.020 mg/L). Thus, the concentration of methane and ethene appeared to increase over the general background in areas where the chlorinated solvent constituency indicated the occurrence of reductive dechlorination.

Stable Carbon Isotopes. During the history of site operations a large number of chlorinated compounds have been manufactured as products or precursors. All chlorinated ethenes observed in the groundwater have the potential to be manufactured products rather than biodegradation products. The stable carbon isotope (^{12}C and ^{13}C) composition of the dissolved VC and ethene were examined to determine if these compounds were derived from chemical manufacturing or biodegradation. Naturally, these isotopes exist in a ratio of 1.1 percent ^{13}C to 98.9 percent ^{12}C. The natural abundance of these two isotopes is expressed as the $\delta^{13}C$ and calculated as shown:

$$\delta^{13}C \text{ (in ‰)} = 1000[(^{13}C/^{12}C) \text{ sample}/(^{13}C/^{12}C) \text{ standard} -1]. \qquad (1)$$

Recent evidence suggests that whereas manufactured VC will have a $\delta^{13}C$ value typical of petroleum hydrocarbons (approximately –28 to –32 ‰), VC produced by biodegradation of more chlorinated ethene parent compounds has a wider range of $\delta^{13}C$ values (typically –50 ‰ to + 10 ‰) (Bloom et al., 2000; Slater et al., 2000).

Petroleum and petroleum-derived compounds typically have a $\delta^{13}C$ of -28 to -32‰. Because chlorinated ethenes are prepared using petroleum-derived carbon, the initial $\delta^{13}C$ for chlorinated ethenes should be close to this range. The $\delta^{13}C$ for PCE and TCE was determined from source area groundwater taken from well S-120. The $\delta^{13}C$ was -34 ‰ for PCE and -31 ‰ for TCE (Table 2). Both values were consistent with ranges measured for pure-phase PCE and TCE (Beneteau et al., 1999). While less in known about the isotopic composition of pure phase VC, source area well S-80 has a $\delta^{13}C$ value for VC of -28 ‰, consistent with reported values for manufactured VC (Bloom et al., 2000; Slater et al., 2000).

In contrast, the $\delta^{13}C$ for VC in down gradient well D-200-1 (5.1‰ and 6.5‰ for duplicate measurements) was extremely enriched for ^{13}C compared to the $\delta^{13}C$ of the chlorinated ethenes and VC in the source area wells (Table 2). The positive values for $\delta^{13}C$ indicate a marked enrichment in the ^{13}C versus the ^{12}C content of the VC. The results suggest that the VC in well D-200-1 has experienced a very high degree of biodegradation compared to the VC in the source area. The positive $\delta^{13}C$-values would only occur by the preferential dechlorination or assimilative biodegradation of isotopically light (^{12}C-enriched) VC compared to isotopically heavy (^{13}C-enriched) VC. This conclusion is supported by the fact that well D-200-1 has the highest VC and ethene concentrations, the highest dissolved chloride, the highest dissolved methane, and among the highest dissolved hydrogen concentrations of all wells sampled, conditions known to support and indicate reductive dechlorination.

Stable Chlorine Isotopes. The preferential release of ^{35}Cl during reductive dechlorination will result in the accumulation of ^{35}Cl in the groundwater so that the $\delta^{37}Cl$-value of the groundwater becomes lighter (more negative) than background values. This phenomenon is demonstrated by the change in $\delta^{37}Cl$ in S-80 ($\delta^{37}Cl$ = -2.72‰) compared to the background or estuarine $\delta^{37}Cl$ of 0.06‰ and -0.15‰, respectively.

As dechlorination proceeds, products such as VC become progressively more ^{35}Cl depleted and more enriched in ^{37}Cl. Therefore, after extensive biodegradation, the resulting VC will have a progressively more enriched (positive) $\delta^{37}Cl$-value. Should the VC migrate into an area that has not been significantly affected by the dechlorination of other chlorinated compounds and continue to be dechlorinated or assimilatively biodegraded, the remaining chlorine atoms will be released into the groundwater. The very positive $\delta^{37}Cl$ in groundwater from D-200-1 (0.47‰ and 0.41‰ for duplicate measurements) suggests that chlorine is being biologically released from isotopically heavy VC. This interpretation of the data is consistent with the isotopically enriched stable carbon isotope composition ($\delta^{13}C$) of VC from the same wells.

Groundwater Geochemistry. The results provided indirect evidence that VC is being biodegraded in the 200-Foot Sand. The dissolved oxygen content of the groundwater was very low in all wells except for well S-200-2 indicating anaerobic conditions. No nitrate, limited ferrous iron, a small amount of sulfide and non-detectable concentrations of sulfate, high concentrations of methane (14 to 122 mg/L within and downgradient of the source area), and dissolved hydrogen concentrations consistent with deeply anaerobic conditions (2.7 to greater than 20 nM), all indicate that the 200-Foot Sand is in a geochemical state that will support reductive dechlorination. The total organic carbon content of the 200-Foot Sand ranged from 56 to 78 mg/L. This concentration is generally considered adequate to support and sustain anaerobic biodegradation (US EPA, 1998). The geochemistry data were applied to the US EPA Strength of Evidence screening protocol (US EPA, 1998). The aquifer system achieved a score indicating adequate to strong evidence for conditions that can support anaerobic biodegradation of chlorinated ethenes.

Biodegradation Rate Estimates for Vinyl Chloride. The typical approach for calculating attenuation rate estimates for chlorinated ethenes involves using changes in concentration along down gradient flow paths represented by at least three wells. An appropriate placement of wells was not available. Well S-80 was representative of the source area groundwater chemistry and geochemistry but it does not lie directly up gradient from D-200-1. D-200-2 and D-200-3 are both down gradient and cross gradient from D-200-1. Thus, they are not suitably positioned to support typical rate calculations.

Attenuation Rate Estimated using the $\delta^{13}C$ Change in VC. The stable isotope data provided an indirect means to estimate the VC attenuation rate. The changes in the $\delta^{13}C$ in VC and ethene between source area well S-80 and down gradient well D-200-1 were used to calculate attenuation rates that should correlate to the VC attenuation rate. The rate calculations assumed:
- The change in $\delta^{13}C$ was equivalent to the change in VC concentration
- The VC plume is at steady state and no other sources of VC are contributing to the plume other than those sources accounted for by well S-80
- A zero-order rate calculation (y=mx+b) with units ‰ per time can be converted to a first order rate constant by multiplying by 0.1 to yield percent per time which is the unit of rate derived from an exponential or first order rate equation ($C_i=C_0e^{-kt}$).

Minimum and maximum horizontal groundwater flow velocities of 0.3 ft/day and 0.6 ft/day were used to establish a range of attenuation rates. The results of rate calculations using the $\delta^{13}C$ for VC and the $\delta^{13}C$ for ethene yielded half-lives ranging from 0.5 to 0.9 years (Table 3).

Attenuation Rate Estimated Using the Change in Ethene Concentration Between the Source Area and Down Gradient Wells. A second approach to estimating the VC attenuation rate used the production of ethene between well S-80 and D-200-1. Ethene is the direct dechlorination product of VC and is only produced through the dechlorination of VC. Therefore, the inverse production rate of ethene should approximate the VC attenuation rate. The ethene production rate was calculated using a first-order rate equation and the concentrations observed in wells S-80 and D-200-1. The half-life of VC ranged from 0.4 to 0.7 years using this method (Table 3).

Biodegradation Rate Estimated Based on the Percent Biodegradation Observed from the Source Area to a Down Gradient Well. A third approach to estimating the VC biodegradation rate employed the relationship between the isotopic composition of the residual VC and the extent of biodegradation. This relationship can be modeled by a simple isotopic model known as the Rayleigh model (Bloom et al., 2000; Slater et al., 2000). The Rayleigh model assumes a constant isotopic preference for ^{12}C versus ^{13}C during biodegradation, referred to as the fractionation factor α. The relationship between the concentration of the residual contaminant (or fraction remaining, f) and the isotopic composition of the residual VC ($\delta^{13}C_b$) can be described by a simple Rayleigh closed system set of equations expressed in δ ‰ notation after Mariotti et al. (1981):

$$(\alpha-1) \ln f = \ln((\delta^{13}C_b/1000 + 1)/(\delta^{13}C_o/1000 + 1)) \qquad (2)$$

where $\delta^{13}C_o$ is the initial isotopic composition of the VC, and $\delta^{13}C_b$ is the isotopic composition of the residual VC.

Published values for the fractionation factor (α) during microbial reductive dechlorination of VC fall in a relatively close range (0.9785 to 0.9734, Bloom et al. 2000; 0.9776, Slater et al. 2000). Assuming that these laboratory-derived fractionation factors can be applied to this site, the extent of biodegradation that has occurred between the source well (S-80) and down gradient well D-200-1 can be estimated using this equation.

Substituting site data and published fraction factors into equation 2 and solving for "f" yields an average VC degradation of 77 percent between S-80 and D-200-1. A simple calculation yields biodegradation half-lives ranging from 0.02 to 0.04 years based on the time, as determined by distance and groundwater flow velocity, required to yield 77 percent biodegradation of the VC between the source area and the down gradient area represented by D-200-1 (Table 3). This rate estimate is less influenced by dispersion, diffusion, and retardation compared to other rate calculations that use change in concentration over time or distance. Although the two applications of the $\delta^{13}C$ data and the ethene production data gave different ranges of rates, the rates are generally consistent with rates observed at other sites where VC has been shown to biodegrade (U. S. EPA, 1998, and Brady et al., 1998).

The results from the investigation provide evidence that dissolved VC is being biodegraded in the 200-Foot Sand aquifer. Chlorinated ethenes are dechlorinated as they migrate vertically and horizontally through fractured clays and WBUs above the 200-Foot Sand. Groundwater in the 200-Foot Sand is impacted with VC which appears to be substantially if not completely biologically degraded based on highly enriched (positive) stable carbon isotope ratios.

TABLE 1. Sample locations, water bearing unit and purpose for sampling.

Sample Location	Water-Bearing Unit (feet)	Purpose for Inclusion
BKG	200	Background
S-80	80	Within the chlorinated ethene source area
S-120	120	Within the chlorinated ethene source area
S-200-1	200	Within the chlorinated ethene source area
S-200-2	200	Within the chlorinated ethene source area
S-200-3	200	Within the chlorinated ethene source area
D-200-1	200	Highest concentration of VC, 500 ft down gradient
D-200-1 (dup)	200	Duplicate sample from well with highest VC conc.
D-200-2	200	1000 Ft. down and cross gradient of high VC area
D-200-3	200	1600 Ft. down and cross gradient of high VC area
EST	surface	Background chloride

TABLE 2. Chlorinated solvent concentration and $\delta^{13}C$ at each location.

Sample Location	PCE (mg/L)	$\delta^{13}C$ (‰)	TCE (mg/L)	$\delta^{13}C$ (‰)	DCE (mg/L)	$\delta^{13}C$ (‰)	VC (mg/L)	$\delta^{13}C$ (‰)	Ethene (mg/L)	$\delta^{13}C$ (‰)
BKG	0.005 U	NA	0.01 U	NA	0.010 U	NA	0.002 U	NA	0.001 U	NA
S-80	4.11	NA	11.3	NA	0.619	NA	0.764	-28	0.076	-34
S-120	0.086	-34	0.060	-31	0.038	NA	0.008	<dl	0.003	-37
S-200-1	1.56	NA	0.49	NA	0.200 U	NA	0.040 U	NA	0.001 U	NA
S-200-2	1.03	NA	0.31	NA	0.200 U	NA	0.040 U	NA	0.001 U	NA
S-200-3	0.100 U	NA	0.13	NA	0.043	NA	0.039	<dl	0.001 U	NA
D-200-1	0.100 U	NA	0.100 U	NA	0.200 U	NA	2.67	5.1	4.82	-43
D-200-1 (dup)	0.100 U	NA	0.100 U	NA	0.200 U	NA	2.15	6.5	8.22	-44
D-200-2	0.005 U	NA	0.01 U	NA	0.010 U	NA	0.002 U	NA	0.001 U	NA
D-200-3	0.005 U	NA	0.01 U	NA	0.010 U	NA	0.002 U	NA	0.001 U	NA
EST	0.005 U	NA	0.01 U	NA	0.010 U	NA	0.002 U	NA	0.001 U	NA

NA, not analyzed
U, not detected above the indicated detection limit
<dl, concentration too low for isotope detection (less than 100 ug/L)
mg/L, milligrams per liter
Error associated with the carbon isotope analysis is ± 0.5‰

TABLE 3. Vinyl chloride attenuation rate estimates.

Parameter	Minimum Flow Velocity	Maximum Flow Velocity
Horizontal Flow Velocity	0.3 ft/day	0.6 ft/day
VC Attenuation Rate Based on $\delta^{13}C$ Change in VC	-0.74 per yr	-1.49 per yr
VC Attenuation Half-life in Years	0.9 yrs	0.5 yrs
VC Attenuation Rate Based on Production of Ethene	-0.98 per yr	-1.95 per yr
VC Attenuation Half-life in Years	0.7 yrs	0.4 yrs
VC Attenuation Rate Based on Percent Biodegradation	-16.9 per yr	-33.7 per yr
VC Attenuation Half-life in Years	0.04 yrs	0.02 yrs

REFERENCES

Aravena, R., K. Beneteau, S. Frape, B. Butler, D. Major, E. Cox. 1998. "Applications of Isotopic Finger-Printing for Biodegradation Studies of Chlorinated Solvents in Groundwater." In G. B. Wickramanayake, R. E. Hinchee (Eds.). *Risk, Resource, and Regulatory Issues: Remediation of Chlorinated and Recalcitrant Compounds*. Battelle Press, Columbus, OH. pp. 67-71.

Beneteau, K.M., R. Aravena, S.K. Frape 1999. "Isotopic characterization of chlorinated solvents – laboratory and field results." *Organic Geochemistry* 30, 739-753.

Bloom, Y., R. Aravena, D. Hunkeler, E. Edwards, S.K. Frape 2000. "Carbon Isotope Fractionation during Microbial Dechlorination of Trichloroethene, cis-1,2-Dichloroethene, and Vinyl Chloride: Implications for Assessment of Natural Attenuation." *Environmental Science and Technology*, 34, 2768-2772.

Brady, P. V., M. V. Brady, and D. J. Borns, 1998. *Natural Attenuation CERCLA, RBCA's and the Future of Environmental Remediation*. Lewis Publishers, Boca Raton, FL.

Mariotti, A., J.C. Germon, P. Hibert, P. Kaiser, R. Letolle, A. Tardieux, P. Tardieux, 1981. "Experimental determination of nitrogen kinetic isotope fractionation: Some principle illustrations for the denitrification and nitrification processes." *Plant and Soil* 62, 413-430.

Pedersen, K. S., Aa. Fredenslund, P. Thomassen, 1989. *Properties of Oils and Natural Gases*. Gulf Publishing Co., Houston, TX.

Sherwood Lollar, B., G. F. Slater, J. Ahad, B. Sleep, J. Spivak, M. Brennan, P. MacKenzie, 1999. "Contrasting Carbon Isotope Fraction during Biodegradation of Trichloroethene and toluene: Implications for Intrinsic Bioremediation." *Organic Geochemistry* 30: 813-820.

Slater, G. F., B. Sherwood Lollar, E. Edwards, B. Sleep, M. Witt, G. M. Klecka, M. R. Harkness, J. L. Spivack, 2000. "Carbon Isotope Fractionionation of Chlorinated Ethenes During Biodegradation: Field Applications." In G. B. Wickramanayake, A. R. Gavaskar, M. E. Kelley (Eds.), *Natural Attenuation Considerations and Case Studies: Remediation of Chlorinated and Recalcitrant Compounds*. Battelle Press, Columbus, OH. pp. 17-24.

U. S. EPA, 1998. *Technical Protocol for Evaluating Natural Attenuation of Chlorinated Solvents in Groundwater*. EPA/600/R-98/128.

ONE YEAR OF MONITORING NATURAL ATTENUATION IN SHALLOW GROUNDWATER

L. G. Stehmeier (NOVA Research Technology Centre, Calgary, AB)
Ryan Hornett and Larry Cooke (NOVA Chemicals Corporation, Joffre, AB)
M. McD. Francis (NOVA Research Technology Centre, Calgary, AB)

ABSTRACT: Monitored natural attenuation has been identified, based on risk analysis, as a feasible option for remediating groundwater and soil at a petrochemical site. Four areas designated A, B, C, and D were sampled in March, June, September, and November 1999 and analyzed for stable isotope ratios of residual contaminants in conjunction with hydrocarbon concentration, microbial numbers, and inorganic electron acceptor concentrations. The premise of using stable isotopes is that during metabolism the residual hydrocarbon will be enriched in ^{13}C, the heavier isotope. During this period, total hydrocarbon concentration increased in the four areas monitored. Benzene the major component of C5+ contaminant was chosen as a benchmark. C5+ (hydrocarbon fraction) degraders were present in the groundwater and increased during the monitoring period at all four locations. Concentrations of soluble iron and manganese, indicators of iron-reducing and manganese-reducing microbial activity were above the mean concentration derived from background wells analyzed over five years. Stable carbon isotope analyses for benzene showed very little fractionation, except for one well, in the residual contaminants outside the analytical error of $0.5°/_{oo}$. In previous work it was determined that fractionation was not measurable until at least 80% of the contaminant had been biodegraded. While the lack of fractionation for benzene suggested natural attenuation did not occur, the increase in hydrocarbon concentration would have obscured any ^{13}C enrichment that had taken place.

INTRODUCTION

In 1997 it was recommended that fractionation of ^{13}C in residual hydrocarbons during biodegradation be developed as a novel monitoring tool for *in situ* bioremediation (Francis *et al.*, 1997). This has been undertaken and studied in the ensuing years (Stehmeier *et al.*, 1999a, b) and is currently being tested in specific areas as part of the research component of a petrochemical site's groundwater monitoring program.

The 1999 monitoring program examined four areas of known near surface groundwater contamination. Fourteen wells were monitored on a quarterly basis for hydrocarbon content, inorganic parameters, hydrocarbon degraders, and stable isotope fractionation.

MATERIALS AND METHODS

Analysis of groundwater headspace for hydrocarbon concentration was performed at the petrochemical site's analytical laboratory using a procedure based on US EPA method 3810. Analyses were done using a gas chromatograph fitted with a flame ionization detector.

Microbial numbers were determined for groundwater at each piezometer. Hydrocarbon degrading microbes were enumerated by the most probable number (MPN) technique. Groundwater (0.5 mL) was added to 4.5 mL of hydrocarbon degrading medium (HDM) in glass vials. The HDM medium contained per liter of distilled water: 1 g K_2HPO_4, 1 g KH_2PO_4, 2 g NH_4NO_3, 0.3 g $MgSO_4.7H_2O$, 0.001 g $CaCl_2.2H_2O$, 0.001 g $FeSO_4.7H_2O$. A carbon source (20 µL of C5+ fraction) was added to each vial and the tubes were incubated at 25°C for 15 days. Inorganics in the groundwater were determined using ICP spectroscopy and standard laboratory methods (APHA, 1998) by Envirotest Laboratories, Edmonton, Alberta.

Compound-specific carbon isotope ratios were determined in the Environmental Isotope Laboratory of the University of Waterloo using gas chromatography combustion isotope-ratio mass spectrometry (GC-C-IRMS). The GC-C-IRMS system consisted of a Hewlett Packard 6890 GC (Agilent, Palo Alto, USA) with a split/splitless injector, a Micromass combustion interface operated at 850°C, a cold trap cooled to −100°C using liquid nitrogen and a Micromass Isochrom isotope-ratio mass spectrometer (Micromass, Manchester, UK). The GC was equipped with a RTX-5 column (60m x 0.25mm, 1 µm stationary phase, Restec, Bellefonte, USA). The following oven temperature program was used: 20°C for 5 min, 2°C/min to 150°C, hold for 4 min, 20°C/min to 220°C, hold for 2 min. The injector was equipped with a split/splitless sleeve (Restec, Bellefonte, USA) and set at a temperature of 270°C. The compounds were extracted from the aqueous phase using a headspace technique. A headspace of 8 ml volume was created by replacing aqueous solution with helium. After shaking the vials (40 ml) for at least 2 hours on a rotary shaker (150 rpm), between 0.05 and 1 ml of gas phase was injected into the GC at a split ratio of 10:1. All $^{13}C/^{12}C$ ratios are reported in the usual delta notation ($\delta^{13}C$) referenced to the VPDB (Vienna Peedee Belemnite) standard. The $\delta^{13}C$ value was defined as $\delta^{13}C = (Rs/Rr - 1) \times 1000$, where Rs and Rr are the $^{13}C/^{12}C$ ratios of the sample and the international standard, respectively. The precision of the measurements was ±0.5‰.

RESULTS

Four locations, A, B, C, D, known to have hydrocarbon contamination, were chosen at the petrochemical site. These sites were monitored in March, June, September, and November 1999. During this period the total hydrocarbon concentration increased at all four sites (Figure 1). Figure 2 presents the C5+ degrading bacterial numbers for the four areas. These results follow a similar trend to that seen in Figure 1. Figures 3A and 3B present the soluble iron and manganese concentrations, respectively, found in the groundwater at the four areas. The Fe^{+2} concentrations in Areas B and D were above the mean concentration (0.32 mg/L Fe^{+2}) derived from background wells analyzed over five years. The Mn^{+2} concentrations were above the background levels (0.15 mg/L Mn^{+2}) in all areas. The stable carbon isotope data for benzene (Figure 4) at these four sites remained constant and did not indicate significant fractionation occurred. In one well, GW340_HS1, (Figure 5) a decrease in hydrocarbon concentration matched by an increase in $\delta^{13}C$ for benzene was observed.

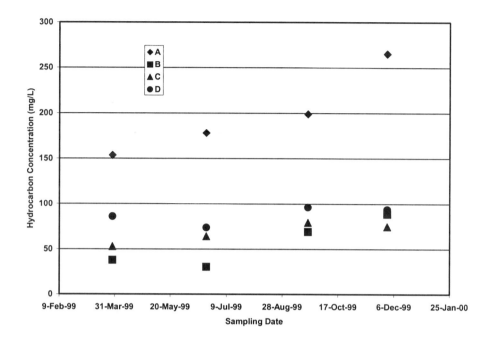

FIGURE 1. Hydrocarbon concentrations observed in four areas of a petrochemical plant during a one-year period.

DISCUSSION

Hydrocarbon concentrations in all four areas increased during the year. The results presented in Figure 1 are the mean of the concentrations for the wells in that particular area. The number of wells analyzed was variable in different areas. The upward trend suggested natural attenuation did not occur during this period at a fast enough rate to maintain or decrease concentrations. No information can be derived from this Figure on the status of contaminant transport. The National Research Council (1993) suggested decrease in hydrocarbon concentration was one of the most important proofs for natural attenuation.

The increase in microbial degraders of the C5+ hydrocarbon fraction during 1999 suggested appropriate microorganisms were present at the site. The increase in numbers followed the increase in concentration suggesting an active microbial population consumed the substrate, as it became available. Additional evidence of microbial activity was found by analyzing for soluble iron and manganese, common electron acceptors in the soil at the petrochemical plant. The above background concentrations of Fe^{+2} and Mn^{+2} indicated microbial metal reduction, however, from the results shown in Figures 3A and 3B no conclusions can be made linking the biodegradation of C5+ with iron and manganese as electron acceptors.

The National Research Council (1993) defined the proof necessary for establishing natural attenuation at a site as loss of hydrocarbons, laboratory assays showing the presence of hydrocarbon degrading bacteria, and field evidence that hydrocarbon biodegradation was occurring. In this work no loss of hydrocarbon was seen but evidence of hydrocarbon degrading bacteria was collected and evidence that hydrocarbon biodegradation had occurred could be indirectly inferred from the increased concentrations of iron and manganese. However, there was no direct evidence linking C5+ degraders with field hydrocarbon degradation.

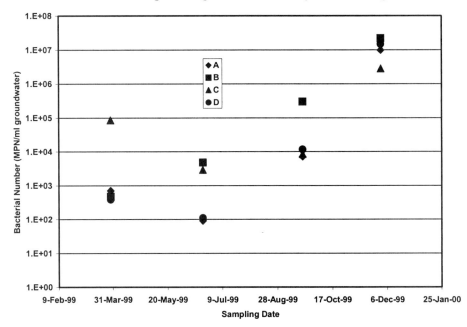

FIGURE 2. C5+ hydrocarbon-degrading bacterial counts determined for four contaminated areas at a petrochemical plant during one year of observation.

The difficulty of linking all three lines of evidence for natural attenuation has led to the use of stable isotopes as a method for monitoring natural attenuation (Francis et al., 1997; Stehmeier et al., 1999a, 1999b). Preferential metabolism of organic compounds containing ^{12}C atoms by microorganisms leads to an enrichment of molecules containing the heavier ^{13}C isotope in the remaining fraction. This enrichment or fractionation occurs because of the higher activation energy required by microorganisms to metabolize compounds containing the ^{13}C isotope. It is established in the literature that the stable isotope fractionation found in nature is due to biological metabolism. Consequently, the residual substrate has

Carbon Isotope Analyses for Monitoring Biotransformation Processes 121

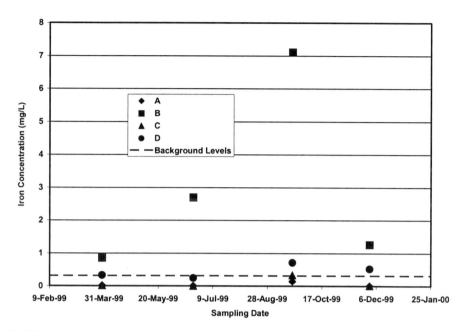

FIGURE 3A. Dissolved iron concentrations analyzed in groundwater from four areas at a petrochemical plant.

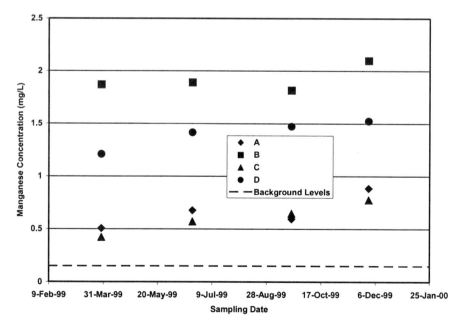

FIGURE 3B. Dissolved manganese concentrations analyzed in groundwater from four areas at a petrochemical plant.

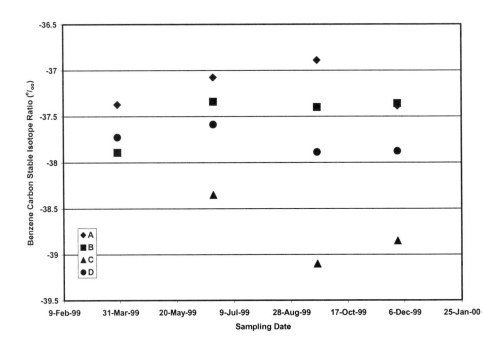

FIGURE 4. Stable isotope ratios of benzene in the piezometers during the sampling period as an indication of natural attenuation.

a heavier isotope ratio than the original substrate. This fractionation can be followed over time resulting in a fractionation factor (Hoefs, 1997). Once a fractionation factor is determined it can be used to predict the extent of degradation. Figure 4 suggests the biodegradation of benzene during the sampling period was minimal. This conclusion is consistent with the general increase in hydrocarbon concentration seen in Figure 1. With increase in hydrocarbon concentration any enrichment in ^{13}C in the residual would be diluted and no fractionation would be seen. Figure 4 represents this scenario.

One well in area A showed the pattern expected when natural attenuation is ongoing. In this well a decrease in C5+ was observed followed by an increase and then a significant decrease. This data occurred over a period of 18 months and is shown in Figure 5. The benzene ^{13}C isotope ratio shown in Figure 5 was the inverse of the hydrocarbon concentration pattern. As concentration decreased, fractionation of ^{13}C occurred and the residual was enriched in ^{13}C. When the hydrocarbon concentration increased, the isotope ratio became more negative indicating natural attenuation was not occurring at a rate sufficient to overcome the dilution of the residual hydrocarbon. The reversal of the stable isotope fractionation observed in December 1999 suggests such a dilution of residual hydrocarbon. After one year the concentration decreased dramatically with a concurrent fractionation of benzene, indicating the loss of benzene was a result of biodegradation. Figure 5 ties the loss of hydrocarbon to biodegradation.

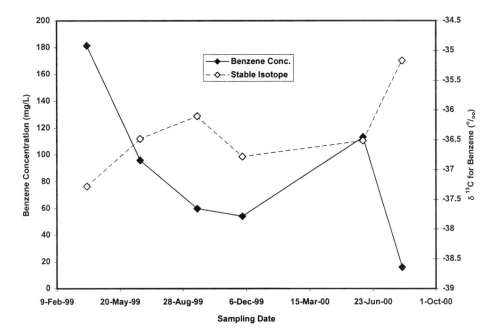

FIGURE 5. Total hydrocarbon loss versus benzene $\delta^{13}C$ at well GW340_HS1 for a sampling period of 17 months.

CONCLUSIONS

- During the sampling period total hydrocarbon generally increased.
- Evidence of bioremediation potential in the form of contaminant degrading bacteria was obtained from all four areas.
- Concentrations of soluble iron and manganese were generally above a 5-year background average, indicating metal reduction as an electron acceptor process during microbial metabolism.
- The benzene ^{13}C ratios remained relatively constant during the sampling period. The lack of fractionation implies significant loss did not occur during the sampling period.
- Well 340-HS1 did show natural attenuation was occurring in its vicinity by having a decrease in total hydrocarbon and an increase in $\delta^{13}C$ benzene values.

REFERENCES

American Public Health Association. 1998. *Standard Methods for the Examination of Water and Wastewater* (20th Edition), ISBN 0-87553-235-7.

Francis, M., L. Stehmeier and R. Krouse. 1997. "Techniques for monitoring intrinsic bioremediation." Paper 97-42, 7 pp., *The Petroleum Society 48th Annual Technical Meeting, June 8 – 11, 1997, Calgary, Alberta, Volume 2.*

Hoefs, J. 1997. *Stable Isotope Geochemistry, 4th Edition*. Springer-Verlag, Berlin. ISBN 3-540-61126-6.

National Research Council. 1993. *In Situ Bioremediation: When Does It Work?* National Academy Press, Washington, D. C. Pp. 64.

Stehmeier, L. G., E. J. M. Diegor, M. McD. Francis, L. Winsor and T. A. Abrajano, Jr. 1999a. Use Of Isotope Fractionation In Residual Hydrocarbons For Monitoring Bioremediation. *Natural Attenuation of Chlorinated Solvents, Petroleum Hydrocarbons, and Other Organic Compounds*, 5(1). ISBN 1-57477-074-8. Pages 207-212.

Stehmeier, L. G., M. McD. Francis, T.R. Jack, T. A. Abrajano, Jr., L. Winsor and E. Diegor. 1999b. Field evidence for in situ bioremediation using compound specific $^{13}C/^{12}C$ ratio monitoring. *Organic Geochemistry* 30(8A): 821-833.

MODEL ANALYSIS OF REDUCTIVE DECHLORINATION WITH DATA FROM CAPE CANAVERAL FIELD SITE

Christine Shoemaker, Matthew Willis, Wei Zhang, James Gossett
School of Civil and Environmental Engineering
Cornell University, Ithaca, N.Y. 14853 U.S.A.
Cas12@cornell.edu

ABSTRACT: A groundwater transport model incorporating the effect of groundwater pumping and hydrogen donor injection on chemical and biological kinetics of reductive dechlorination is applied to a set of field data collected at Cape Canaveral. Kinetic parameter values used are identical to those from a microcosm study for this site.

INTRODUCTION

An important class of groundwater contaminants consists of chloroethenes—tetrachloroethene (PCE) and its daughter products: trichloroethene (TCE), dichloroethene (DCE) and vinyl chloride (VC). These materials can be remediated *in situ* by anaerobic biodegradation to non-toxic ethene. Injection of organic compounds (donors) that ferment to provide hydrogen can enhance the activity and growth of dechlorinating microorganisms that degrade chloroethenes.

Incorporating the interactions observed in Fennell et al (1997), Fennell and Gossett (1998) developed a biokinetic model employing Michaelis-Menten-type kinetics to examine the fermentation of electron donors and competition for the evolved H_2 between hydrogenotrophic dechlorinators and methanogens. Special features of the model include H_2 thresholds (below which no activity occurs) that are lower for dechlorinators than for methanogens. Also included are thermodynamic hydrogen ceilings on donor fermentation for specific donors (butyric acid, ethanol, lactic acid, and propionic acid). The model includes the effects of hydrogen and substrate on the growth rates of both dechlorinating and methanogenic organisms in order to describe accurately the competitive processes. Model simulations compared favorably to experimental data for the four different donors. The Fennell and Gossett model is a batch model (of a small, completely mixed reactor) that ignores transport of contaminants, spatial variability, and the effects of pumping.

Shoemaker and Willis (Willis et al., 1999; Willis and Shoemaker, 2000; Willis 2000) have developed a three dimensional fate-and-transport model for halogenated-organic contaminants that properly incorporates hydrogen dynamics and competition. Unlike the earlier batch model, the Willis and Shoemaker model describes the movement of contaminants, donors and microbes through space. The dechlorination transport model builds upon the publicly available transport codes MT3D (Zhang and Wang, 1998) and RT3D (Clement et al., 1998). The Willis and Shoemaker model uses the reaction kinetics equations from the batch model (which has no transport) that was developed and verified with laboratory data by Fennell and Gossett (1998 and Fennell et al., 1997).

OBJECTIVE

This paper will describe the comparison of the Willis and Shoemaker dechlorination transport model predictions to field study site data collected at Cape Canveral, Florida. The purpose of this analysis is to use the combination of field measurements and model to better understand the dynamics of the processes at the field site and to assess the predicitive capabilities of the model.

MODEL

In addition to the reactions incorporated in the Fennell and Gossett batch model, Willis and Shoemaker (2000) have imbedded the following in the groundwater transport model: a) movement of all the constituents with the flow of the groundwater, and b) partitioning of the constituents onto aquifer material, c) the effects of pumping on transport. The version of the model used in the Cape Canaveral analysis is three dimensional and incorporates recirculating wells, which can increase contact between dechlorinators and chlorinated ethenes and thereby speed up bioremediation.

FIELD DATA

The data used in this analysis is from a 1999 field study (Alleman et al, 2000) performed at Cape Canaveral Air Station, Florida, which is an implementation of the reductive anaerobic biological in situ treatment technology (RABITT) treatability test protocol. The data collected include contaminants and daughter products (TCE, DCE, vinyl chloride), geochemical parameters and measurements to estimate groundwater velocity and direction. (PCE is not present at this site.) The data also includes the results of injections of bromide, a conservative tracer used to determine the velocity field of the groundwater when the injection and extraction wells are operating. The field site used a closely spaced multiple point recirculating injection-extraction well design with a subsurface admixture of lactate as a donor to supply hydrogen for the dechlorination of TCE. There are 43 sampling points located in different depths and horizontal locations.

COMPARISON OF MODEL RESULTS TO FIELD DATA

The bromide information was used to calibrate the hydraulic parameters in the model, including hydraulic conductivity, porosity, and boundary conditions. The full model with all the chemical reactions was then run using the parameter values determined in the microcosm experiments by Fennell and Gossett. (The temperature at the site was $24°C$ and the temperature in the original laboratory experiments was $35°C$.

Figure 1 shows the initial concentrations of TCE at the site and Figure 2 shows the results 576 hours (24 days) later after the continuous injection of lactate and water. These views are from a vertical slice going down the middle of the site, with the natural gradient flow of groundwater going from left to right. In Figure 1, the iso-concentration contour curves represent the initial concentration input to the model. This input was determined by fitting a three-dimensional spline (a smoothing function) through the initial observations at 43 points in three dimensions. The spline was allowed to interpolate within the spatial region for which there was data. Outside this interpolation region, the initial concentration was assumed to match that at the boundary of the interpolation area, which is why the iso-concentration lines are straight in Fig. 1 outside the interpolation area.

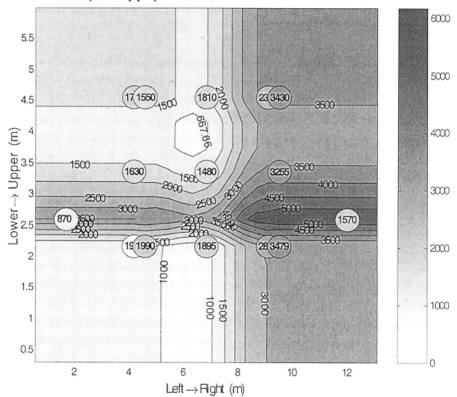

FIGURE 1. Initial concentrations of TCE at Cape Canaveral field study plot. This plot is a vertical slice through the middle of the field site. The numbers in ovals indicate the concentrations measured at the specified locations. The contour lines of iso-concentration curves describes the smooth distribution of concentrations that was used as initial concentrations for the 2000 nodes in the simulation model. The grey scale indicates the concentration area. The purpose of including the area outside the interpolation area in the simulation was to incorporate the mixing that occurred outside the sampled area.

The iso-concentration curves in Fig. 2 are the results of the model simulations. In this case the numerical values of the model predictions of the concentrations at 3289 node points in the model simulation are smoothed out by a software program to obtain the iso-concentration curves describing the surface. The concentrations corresponding to each iso-concentration line is printed on the graph. The ovals denote the observed data at the corresponding location. The number inside the oval is the concentration from the observed data.

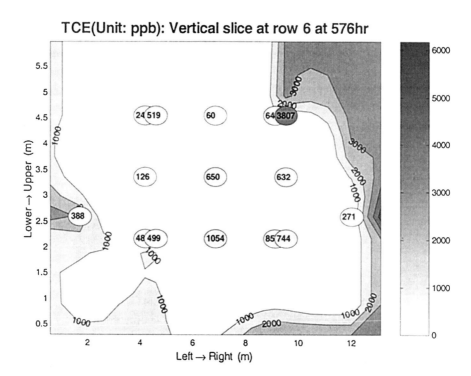

FIGURE 2 The ovals indicate the measured concentrations 576 hours after the start of lactate injection in a vertical slice through the middle of the field study site. The contour iso-concentration lines give the smooth surface that fits the model-predicted values at the 3289 nodes in the simulation model.

CONCLUSIONS

The results presented in Figure 2 are based on the kinetic parameter values from a microcosm study done by Fennell and Gossett (personal communication) and these values are close to those published in the earlier paper by Fennell and Gossett (1998). There was no calibration of the kinetic parameter to obtain these results, which supports not only the validity of the model but also the values of the kinetic parameters. We would expect that at other sites the kinetic parameters might require some adjustment, but these results do indicate that the microcosm parameter set is a good initial guess of parameter values for sites with the same active organisms and similar site conditions. The hydrologic parameters were calibrated (adjusted) to fit the bromide data before the chemical model results were obtained to insure that the velocity field in the field study plot was adequately represented; the hydrologic parameters were not changed to improve the fit of the observed values of the chlorinated ethenes.

FUTURE WORK

We will continue to do additional comparisons of model simulations to observed data to assess the spatial and temporal changes in concentrations. The simulations appearing in this paper use kinetic parameters that fit microcosm data collected by Gossett and Fennel (personal communication). We will assess the differences in model predictions when using other parameter values including those that appeared in the original batch model paper by Fennell and Gossett (1998).

This effort is part of a larger effort to develop modeling tools that can be used to help environmental consultants design effective remediation plans for dechlorination. The model can incorporate site specific conditions (by inputing appropriate data for conductivities, porosity, etc.) and can be used to evaluate the likely difference in clean-up efficiency associated with alternative locations of injection and extraction wells and the rates of donor injection (by varying the location of the wells and injection rates in each of a number of runs of the simulation model). There is a related poster at this Battelle Conference on the current development of DECHLOR, a user friendly, publicly available version of a model using the equations discussed above.

ACKNOWLEDGEMENTS

This study is funded by the N.Y. State Advanced Technology Program in Biotechnology through a grant to Professor Shoemaker and Professor Gossett. Cornell University. This study is done in co-operation with HydroMath, LLC, which is developing a user-friendly model DECHLOR of the effects of engineered dechlorination. The data used in the analysis were collected in the study described by Alleman et al. (2000), which was sponsored by ESTCP in DOD. We appreciate the efforts of B. Alleman and J. Morse to provide and explain the Cape Canaveral field data. R. Clement of DOE has given us access to beta versions of RT3D codes that have improved our numerical results (Clement et al., 1998).

REFERENCES

Alleman, B. E.; J.M. Morse, F. Snyder, L. Ackert, G. Sewell, J.M. Gossett, D. E. Fennell, " Reductive anaerobic biological in situ treatment technology (RABITT) treatability test: Results from Cape Canaveral Air Station", Presented at the Second International Conference on Remediation of Chlorinated and Recalcitrant Compounds. Monterey, CA, May 22-25, 2000.

Clement T. P., Y. Sun, B.S. Hooker and J.N. Petersen, "Modeling multispecies reactive transport in groundwater", *Ground Water Monitoring and Remediation*, 18, 79-92 ,1998.

Fennell, D. E., J. M. Gossett, and S. H. Zinder, "Comparison of Butyric Acid, Ethanol, Lactic Acid, and Propionic Acid as Hydrogen Donors for the Reductive Dechlorination of Tetrachloroethene," *Environmental Science & Technology*, 31, 918-926 ,1997.

Fennell, D. E., and J. M. Gossett, "Modeling the Production of and Competition for Hydrogen in a Dechlorinating Culture," *Environmental Science & Technology*, 32, 2450-2460, 1998.

Willis, W., C. Shoemaker, J. Gossett, D. Fennell, "Applications of a Competitive Hydrogenotrophic Biological Dechlorination Transport Model for Groundwater Remediation," in *Engineered Approaches for In Situ Bioremediation for Chlorinated Solvent Contamination 5(2) 27-33*, Battelle Press, 1999

Willis, W. and C. Shoemaker, "Engineered PCE Dechlorination Incorporating Competitive Biokinetics: Optimization and Transport Modeling" in *Bioremediation and Phytoremediation of Chlorinated and Recalcitrant Compounds, 2(4) 311-318,* Battelle Press, 2000.

Willis, M. B. *Modeling, Optimization and Sensitivity Analysis of Reductive Dechlorination of Chlorinated Ethenes With Microbial Competititon in Groundwater*, Ph.D. Thesis, School of Civil and Environmental Engineering, Cornell University, May, 2000.

Zheng, C. and P. Wang, "MT3D, A Modular Three-Dimensional Multispecies Transport Model, Documentation and User's Guide" ,1998.

BIOREMEDIATION MODELING: FROM THE PILOT PLANT TO THE FIELD

Marco Villani, Marco Padovani, Massimo Andretta, Mariangela Mazzanti, *Roberto Serra* (Centro Ricerche Ambientali - Marina di Ravenna, Ravenna, Italy)
Beate Mueller, Hans Peter Ratzke (Umweltschutz Nord, Ganderkesee, Germany)
Rocco Rongo, William Spataro, Salvatore Di Gregorio (Università della Calabria, Arcavacata, Cosenza, Italy)

ABSTRACT: A new cellular automata model of the complex set of interacting phenomena which take place in bioremediation is described. The model allows the scaling between pilot plant situations and field operations. The general framework and some transition functions are shown, and an example including (fluid dynamical, chemical and biological layer) is analyzed and compared with experimental results. It is shown, in two real cases, that the model can provide a good forecast of the kinetics in the field, outperforming conventional methods.

INTRODUCTION

In this paper we present a mathematical model based on the discrete framework of cellular automata, which describes the process of bioremediation of contaminated soils. To improve the effectiveness and to widen the applications of in situ bioremediation there is a strong need for dynamical models that support the understanding of the relevant phenomena and allow effective forecasting of the duration of the intervention. In this work we show, in two real-world cases, that models can support the scaling operation, from the pilot plant to the field scale.

BASIC FEATURES OF THE MODEL

The cellular automata (or CA) approach appears particularly well suited to deal with the interacting physical, chemical, biological phenomena that may affect bioremediation (Serra et al., 1998). In order to model the complex phenomenon of the bioremediation it is useful to utilize a macroscopic CA approach (Di Gregorio and Serra, 1999).

IN CA the space is divided into discrete cells, which will be called here "sites" (CAS) in order to avoid confusion with bacterial cells. Each Cellular Automata Site is associated to several substates that represent different characteristics (considered constant inside the CAS) of the space occupied by the site. The substate value can vary depending on interactions among substates inside the CAS (internal transformation) and/or local interactions among CASs. The finite discretization of the space and of the time allows, when possible, the use of a closed explicit solution of the internal transformations that, otherwise, have to be described in terms of ordinary differential equations. A more detailed discussion about the intrinsic characteristics of the Cellular Automata and their peculiarities can be found in (Di Gregorio et al., 1997).

The bioremediation model describes the fluid dynamical properties (multiphase flow in a porous medium), the fate of chemicals (advection, dispersion, sorption, chemical reactions), the growth of the biomass and its interaction with nutrients and contaminants.

Using a formal representation, we can define the general Cellular Automata model *CabCol* as:

$$CabCol = (R, X, Q, P, \sigma, I, \Gamma)$$

Where:

- $R = \{(x, y, z) \mid x, y, z \in N,\}$ is the set of points with integer co-ordinates in the finite region where the phenomenon evolves.
- X identifies the geometrical pattern of CASs, which influence the CAS state change. They are the CAS itself and the "up", "north", "east", "west", "south" and "down" neighbouring CASs:
- Q is the finite set of states of the CA
- $\sigma: Q^7 \to Q$ is the deterministic state transition function for the CASs in R.
- I is set of CASs, to which special conditions are imposed (e.g. those which describe pumps).
- Γ is the set of opportune transformations of the transition function in order to model bondary conditions or external inputs.

The generalized Fluid dynamical properties. The multiphase flow in a porous medium is described by Darcy law:

$$q_\alpha = -\frac{k_{r\alpha} k}{\mu_\alpha} \nabla (p_\alpha + \rho_\alpha gz) \qquad (1)$$

where q_α (m/s) is the volumetric flow, k (m^2) is the permeability of the solid matrix, $k_{r\alpha}$ is the relative permeability of the phase α, m_a (Pa s), p_α (Pa) and ρ_α (kg/m^3) are the viscosity, pressure and density of phase α respectively. $k_{r\alpha}$ takes into account the presence of different phases in the soil and depends upon the saturation S (defined as the ratio between the volume of the phase α and the pore volume in a reference volume) (Bear, 1979); g is the gravitational acceleration and z the altitude.

For a clearer presentation, we can limit ourselves to two-phase systems: the scheme for three-phase systems is similar. In a two-phase system there is a correlation between the pressure of the wetting phase (p_w) and the non-wetting phase (p_{nw}), expressed in terms of the capillary pressure p_{cnw} (the pressure discontinuity at the interface of the two phases); the capillary pressure can be expressed in terms of the saturation of the wetting phase S_w (Leverett, 1941; Van Genuchten, 1980):

$$p_{nw} = p_w + p_c(S_w); \qquad p_c(S_w) = \frac{1}{\alpha}\left(S_e^{-1/m} - 1\right)^{1/n}, p_c > 0 \qquad (2)$$

The relative permeability depends on the presence of the different phases, too. In a two-phase system the relative permeability of each phase can be expressed in terms of the effective saturation S_e of the wetting phase (Mualem, 1976).

$$k_{rw} = S_e^{1/2}[1-(1-S_e^{1/m})^m]^2 \quad (3a)$$

$$k_{rn} = (1-S_e)^{1/2}[(1-S_e^{1/m})^m]^2 \quad \text{with: } S_e = \frac{S_w - S_{wr}}{1 - S_{wr}} \quad (3b)$$

The water pressure is taken as reference pressure: the pressures of the other phases are obtained by applying relation (2). The information inside the physical system is transmitted by means of the potential field; its changes propagate faster than phase flows so we can suppose that phases moves only after that the potential field has reached a stationary state (adiabatic approximation).

Particularly, mass conservation requires that the algebraic sum of the incoming flows (f_k^i) and the outgoing flows (f_k^o) in a CAS vanish:

$$\sum_{k=1}^{6}(f_k^i + f_k^o) = 0 \quad (4)$$

By inserting equation (1) and (2) in (4) (and assuming, for simplicity, cubic CASs of $cell\_size$ size), we can derive:

$$\sum_{j=1}^{6}\left(\frac{k_0^{rw}k}{\mu_w}\frac{\Delta p_j^w + \rho_w g \Delta z_j}{cell\_size} + \frac{k_0^{ra}k}{\mu_a}\frac{(\Delta p_j^w + \Delta p_j^{cwa}) + \rho_a g \Delta z_j}{cell\_size}\right) = 0 \quad (5)$$

with: $\Delta p_j^w = p_j^w - p_0^w$ $\Delta z_j = z_j - z_0$
where the subscripts w and a means, respectively, water and air; p is the pressure, ρ the density, μ the viscosity. From this equation we can derive the value of the water pressure of the central CAS as a function of the pressures of water in the neighboring CASs. This transition function is iterated over the whole cellular automata until a stationary condition is reached; from this potential distribution we can compute the phase flows, which follow the gradient of the potential.

Fate of chemicals. The most relevant phenomena that we model are the transport of a chemical due to the motion of the phase in which it is present, the molecular diffusion of a chemical within a phase, the transport of a chemical among phases and chemical reactions. Transport and diffusion take place between different CASs, while adsorption/desorption and chemical reactions are supposed to occur inside each CAS.

We suppose that a chemical transported by a mobile phase mixes instantaneously with the chemical already present in the CAS. The variation in the amount of chemical within the CAS is given by the difference between the chemical that comes into the CAS and the chemical that flows out from the CAS.

The amount of chemical Q_C ([m]) in phase α that in a step of CA flows in (or out) the CAS in the direction i is:

$$Q_C = J_{\alpha,i} \cdot C_i \qquad (6)$$

where $J_{\alpha,i}$ is the flow ([$l^3 t^{-1}$]) of the phase α in the direction i in the CAS and C_i is the concentration of the chemical in the CAS i of the neighbor ([m l^{-3}]).

Mass transfer driven by the difference of concentration of two neighboring CASs is modeled as follow:

$$Q_C = -(D_\alpha + \lambda_\alpha v) \cdot \frac{\Delta C}{\Delta z} \cdot A \qquad (7)$$

where D_α is the diffusion coefficient ([$l^2 t^{-1}$]) in phase α, λ_α is the dispersivity, v is the average fluid velocity (Jury, 1991), and C is the concentration of the chemical in the CAS ([m l^{-3}]) and A ([l^2]) is the section perpendicular to the flow (Bear, 1979).

Flow among phases and adsorption/desorption (when one of the two phases involved is the soil) can be modeled in the same formal way. The amount of chemical Q_C ([m]) that moves from phase α to phase b in one step on the CA is given by:

$$Q_C = Q_{oC} \exp(-\Gamma_{\alpha \to \beta} \cdot \Delta t) \qquad (8)$$

where Q_{0C} ([m]) is the amount of chemical C at the previous step, $\Gamma_{\alpha \to \beta}$ ([t^{-1}]) is the exchange parameter among phase α and β and Δt is the time corresponding to one step on the CA. Notice that eq. (8) is directly the solution of a differential equation.

Other internal phenomena involve chemical reactions, which can be described using a common scheme, in which chemical species combine in quantities related to their stoichiometric coefficients. An important example is the dissociation of hydrogen peroxide in oxygen and water that has been modeled considering that the amount of oxygen produced by the dissociation, $Q(O_2)$, is related to the amount of hydrogen peroxide through their stoichiometric coefficients: $Q(O_2)=16/34 \; Q(H_2O_2)$.

Biomass growth. Many different processes lead to the biodegradation of an organic compound. In the application of the model described in this work the main factors that describe the behavior of a bacterial population are the spontaneous growth and death, the decrease due to the presence of poisonous chemicals and the growth due to the degradation of a specific compound.

If we have a single chemical which is degraded by a single bacterial population, we simulate the variation of the concentration of the bacterial population X during a time step by the following equation:

$$\Delta X = aX - bX^2 + \mu_{max} \frac{C_{O_2}}{k_{O_2} + C_{O_2}} \cdot \frac{C}{k+C} \cdot X - K \cdot C \cdot X \qquad (9)$$

The first two terms are the Verhulst terms. The first term describes the linear growth of the biomass, the second one describes the microbial endogenous decay due to high biomass concentration. The third term of (9) describes the microbial growth in terms of Monod-like terms product (Thullner and Schaefer, 1999); the first Monod-like term describes the effect of the oxygen concentration (C_{O2}), the second one the effect and the chemical concentration. These two terms can be separately determined by experiment. The potential poisonous effect of the chemical is described by the fourth term of (9).

The variation of the chemical during a time step is described through a Monod-like oxygen limited decrease term:

$$\Delta C = -Y\mu_{mai} \frac{C_{O_2}}{k_{O_2} + C_{O_2}} \frac{C}{k+C} \cdot X \qquad (10)$$

SCALING

While adopting the CA approach has proven effective in terms of speed of development and efficiency of implementation on parallel computers, we believe that the major strength of the model derives from the method that is described in this paper, which is divided into two major phases. The first phase takes place in the laboratory and is based on the use of pilot plant filled with the soil coming from the real site; the dimensions of pilot plants may differ, but typical linear dimensions do not exceed 1-2 meters. In this phase a general model of the process is tailored to the contaminant, soil and bacterial characteristics of the case at hand. Tailoring the model can be viewed as a learning process, which involves the human researcher and is supported by computer-based tools: the outcome of this phase is both a better understanding of the most relevant processes which occur, and a "minimal model" which describes these most relevant processes.

The second phase involves forecasting the results of the field intervention. In case there were significant differences between these forecasts and the observed behavior, a further learning phase would be required.

We tested this approach in the Colombo project (Esprit n° 24907), by simulating two real bioremediation interventions and the related pilot plants.

The pilot plants. The same procedure is used for both situations: therefore, in the following we present the pilot plant 1 (related to the field test 1) in a deeper way; very similar results hold for the pilot plant 2 (related to the field test 2).

In order to predict the outcomes of the field operations, for each experiment two variants of the pilot tests have been performed. In the first variant of the experiments we used water with nutrients and in the second variant we used water enriched with nutrients and with hydrogen peroxide. The plants are composed by three tubes, each one with a diameter of 10 cm and an overall length of 50 cm, arranged into a pile and connected by pipes where the extraction of water samples is possible.

For model testing, we performed an optimization procedure (Di Gregorio et al, 1997), using data obtained from Tube 1 of the Variants 1 and 2 as training set, and the experimental results of Tube 2 of the same variants as test set.

In the optimization process we also suppose the nutrients in the soil don't represent a limiting factor for biomass growth. Finally, following the bioremediation scheme of Christensen and McCarty (Christensen and McCarty, 1975), we suppose that the most probable ratio between the contributions of biomass growth and respiration processes is about 70:30, a value already used in the literature (Nicol et al., 1994). In this way, there are only a few variables, all related to bacterial kinetic coefficients, which are not determined and which have to be optimized. Figure 1 shows the simulation results.

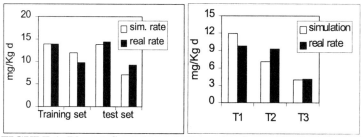

FIGURE 1. Plant 1 final degradation rates. Left side: training set (tubes 1 of Variant 1 and 2) and test set (tubes 2 of Variant 1 and 2). Right side: the rate profile of the three tubes of Variant 1

Biological reactions that don't utilize oxygen as an electron acceptor when the oxygen concentration is below a required threshold have been simulated by means of a first order kinetics. In our pilot plant experiments we found this condition in the third tube of Variant 1 (Pilot Plant 1). In this case, a poor injection of oxygen and very low values of redox potential indicate that here non aerobic phenomena took place. The estimate of the contribution of the aerobic and anaerobic bacteria to the bioremediation inside Tube 3 of Plant 1 is 13.5% and 86.5% respectively.

The field tests. The Colombo field test 1 (polluted with heating oil) is a regular hexagon 20 meters of side; on each vertex and at the center there is an extraction well. Inside each triangle, which has the center and two neighboring edges as its vertex, there are three injection wells; all these wells are working simultaneously following a timetable which involves a flow rate change each week. The water table is about at 9.0 m. below ground level, and it is approximately 2.0 m. deep; above the water table the soil is not saturated.

The Colombo field test 2 (polluted with PAH) is an irregular region of about 400 m^2. On two sides two impermeable walls that reach the deep impermeable zone below the water table bound it; the other two sides are open. The water table is at about 2.0 m. below ground level, and it is approximately 16.0 m. deep; above the water table the soil is not saturated.

The model forecasts the experimental data using, for each field test, the parameters which had been determined in the pilot plant, and the fluid dynamic data describing the real operations. In the following, we compare the results from pilot plants, field tests and model simulations schematizing the contaminant degradation data obtained with these three methods as a first order kinetics process (Suarez and Rifai, 1999), using initial and final values. In such a way it is possible to define a time decay constant, characteristic for each system (real field, model, pilot plant). The following table summarizes the result of this analysis.

TABLE 1. Comparison among real field (Field), simulation (Sim.) and typical pilot plant (Pilot plant) decay constants of the two interventions. For each system the typical time decay constant (first column), and its inverse (second column), is shown.

	Time decay constant (day^{-1})			Characteristic decay time (day)		
	Field	Sim.	Pilot plant	Field	Sim.	Pilot plant
Test 1	2.10e^{-03}	1.58e^{-03}	3.08e^{-03}	475	634	324
Test 2	3.24e^{-03}	3.44e^{-03}	2.52e^{-02}	308	290	40

The time behavior of the contaminant of the two field tests is very different: the model is able to capture this particular feature, while a forecast made using the pilot plants data would fail for one order of magnitude in Test 2. The main cause of this difference is the transport phenomenon, which is very important during the real field operations and explicitly modeled by the model.

CONCLUSIONS

The satisfactory prediction of the field test results is the main result of this work, which shows that it is possible to forecast correctly the duration of intervention (and therefore to have better estimates of the costs of intervention) starting from the data of suitable pilot plants, by using a model for the scaling procedure. As a limited number of tests have been performed so far, further tests are required in order to thoroughly assess the range of validity of the model.

REFERENCES

Bear J. 1979. *Hydraulics of groundwater*. New York: McGraw Hill

Christensen, D.R. & McCarty, P.L. 1975. "Multi-process biological treatment model." *J. Water Pollut. Control Fed.*; **47** (11): 2652-2664.

Di Gregorio, S., Serra, R. & Villani, M. 1997. "A cellular automata model of soil bioremediation." *Complex systems*, **11** (1): 31-54.

Di Gregorio, S. & Serra, R. 1999. "An empirical method for modelling and simulating some complex phenomena by cellular automata." *Future Generation Computer Systems* **16**: 259-271.

Jury, W.A., Gardner W.R., Gardner W.H., 1991. *Soil Physics*. John Wiley & Sons, Inc.

Leverett, M.C. 1941. "Capillary Behaviour in Porous Solids." *Transaction of the AIME*; **142**: 152-169.

Mualem, Y. 1976. "A new model for predicting the hydraulic conductivity of unsaturated porous media." *Water Res. Research*; **12**: 513-522.

Nicol, J.P., Wise, W.R., Molz, F.J. & Benefield, L.D. 1994. "Modeling biodegradation of residual petroleum in a saturated porous column." *Water Resour. Res.*; **30** (12): 3313-3325.

Serra, R., Di Gregorio, S., Villani, M & Andretta, M. 1998. "Bioremediation simulation models." *Biotechnology for soil remediation* R.Serra (ed): Cipa Editore, Milano.

Suarez, M.P., Rifai, H.S. 1999. "Biodegradation Rates for Fuel Hydrocarbons and Chlorinated Solvents in Groundwater". *Bioremediation Journal* 3 (4):337-362.

Van Genuchten, M.T. 1980. "A Closed-Form Equation for Prediction the Hydraulic Conductivity of Unsaturated Soils." *Soil Sci. Soc. Am. J.*; **44**: 892-898.
Thullner, M., Schaefer, W. 1999. "Modeling of a Field Experiment on Bioremediation of Chlorobenzenes in Groundwater". *Bioremediation Journal* 3 (3):247-267.

MODELING THE DYNAMICS OF BIODEGRADATION IN UNSATURATED SOIL FOR IMPLEMENTATION OF ADVANCED CONTROL STRATEGIES

Schoefs O., Chapuis R. P., Perrier M. and Samson R.
(École Polytechnique de Montréal, Montreal, Quebec, Canada)

ABSTRACT : The aim of our study is to optimize bioprocesses in unsaturated soil using advanced control strategies. This work deals with modeling the dynamics of biodegradation in unsaturated soil. First, a hydrodynamic model, describing simultaneous unidirectional water and air flow, based on generalized Darcy's law, was calibrated using a 1.5 m high column of non contaminated soil. The column was equipped with an irrigation and aeration system capable of providing co-current and counter-current water and air flow. Time Domain Reflectometry probes for water content measurement and tensiometers for water and air pressure measurement were placed at six different depths. Second, a biokinetic model, based on oxygen transfer, contaminant transfer and biodegradation rates, was calibrated using data generated from respirometric tests. This provided carbon dioxide production, contaminant consumption and biomass evolution. The column hydrodynamic tests showed that co-current or counter-current water and air flow could significantly influence water infiltration. The Darcy's law-based model described correctly water percolation without air convection. Results from biometric flask experiments indicated that, whatever the water content, the total heterotrophic population remained constant (5.10^7 MPN index/g dry soil), while the hydrocarbon degraders became dominant after only about fifteen days, increasing from 10^5 to 5.10^7 MPN index/g dry soil. After a three-day lag phase, the biometric flask having the lowest water content provided the highest hexadecane mineralization. These results indicate that water content influences hexadecane mineralization especially within the early weeks during which oxygen is probably the limiting factor. After this period, the mineralization seems to be controlled by the hexadecane transfer rate and by the endogenous decay rate.

INTRODUCTION

In situ and on-site bioremediation processes used in unsaturated soil are most often employed under sub-optimal operating conditions, which can lead to an increase in the treatment duration and operating costs, and sometimes to incomplete decontamination. Until now, the lack of reliable biodegradation sensors and the complexity of phenomena involved in unsaturated soil made the implementation of conventional control strategies difficult.

The first difficulty encountered with optimization of biotreatment technologies is controlling water percolation with forced air flow. An undesired water content gradient can be established within the soil, due to a hydraulic gradient caused by the air convection. In some cases, air convection prevents water from infiltrating soils and leads to a local drying or, on the contrary, to local saturation of the soil. Many biotreatment technologies are confronted with this problem such as *in situ* bioventing, above ground biopiles, and biofiltration processes. Transient flow of water into a non-swelling unsaturated soil is well understood and has been largely studied in the literature (Bear, 1972; Fredlund and Rahardjo, 1993; Haverkamp et al., 1977; McDougall and Pyrah, 1998). Nevertheless, prediction using mathematical models is still confronted to two important features: boundary conditions, which are usually not constant, and prediction of hydraulic conductivity in unsaturated soil. To our knowledge, no study has been conducted to predict and validate the influence of forced air convection on transient water flow in unsaturated soils. In this paper, the influence of forced air flow on water percolation is characterized and a mathematical model based on Darcy's law is developed and compared to experimental data.

The second difficulty encountered with optimization of biotreatment technologies is to establish the optimal conditions for biodegradation in unsaturated soil, especially for water content and air flow. Optimal values for these two parameters are difficult to predict because they influence each other. Although high water content is desired to wet soil particle surfaces and hence, lead to increase contact between contaminant and micro-organisms, it can also lead to the development of anaerobic zones or, to a lesser extent, a significant reduction in the oxygen transfer rate which may become the biodegradation limiting factor. In the literature, optimal water content is characterized by relatively large ranges: 25-85% (Sims et al., 1990), 30-90% (Dibble and Bartha, 1979), 40-50% (Cioffi et al. 1991). Therefore a better understanding of the phenomena involved during biodegradation could allow operation of biotreatment technologies under optimal conditions. This paper focuses on the modeling of the three major phenomena that control biodegradation in unsaturated soil: oxygen transfer, contaminant transfer and biodegradation itself.

THEORETICAL BACKGROUND

Transient Flow in Porous Media. Transient flow of a fluid through a porous media is commonly described by Darcy's law :

$$\upsilon = -k\frac{\partial h}{\partial z} \tag{1}$$

where υ is the flow rate of the fluid, k is the permeability coefficient and $\partial h/\partial z$ is the hydraulic head gradient in the z-direction. Darcy's law, when applied to air and water and combined with the continuity equations, leads to the two following partial differential equations:

$$C(\psi)\frac{\partial \psi}{\partial t} = \frac{\partial}{\partial z}\left[K_w(\psi)\left(\frac{\partial \psi}{\partial z} - 1\right)\right] \tag{2}$$

where $\psi = \frac{P_w - P}{\gamma_w}$ (3)

$$\frac{\partial P}{\partial t} = \frac{1}{2}\frac{\partial}{\partial z}\left[K_a(\theta_w)\frac{\partial P^2}{\partial z}\right] \qquad (4)$$

where Ψ is the soil suction, K_w and K_a are respectively the water and air conductivities, P_w and P are respectively the water and the air pressures, γ_w is the volumetric weight of water, θ_a and θ_w are the volumetric air and water contents, and C is the specific water capacity defined as:

$$C(\psi) = \frac{d\theta_w}{d\psi} \qquad (5)$$

Functions $C(\psi)$ and $K_w(\psi)$ can be determined from the soil-water retention curve obtained experimentally. Van Genuchten (Van Genuchten, 1980) proposed the following equations for this determination:

$$\theta_e = \left[\frac{1}{1+|\alpha\psi|^n}\right]^m \qquad (6)$$

where θ_e is the degree of saturation, and α, m and n are constants.

$$K_w = K_{sat}\sqrt{\theta_e}\left[1-\left(1-\theta_e^{1/m}\right)^m\right]^2 \qquad (7)$$

where K_{sat} is the saturated hydraulic conductivity.

Biodegradation in Porous Media. Three phenomena occur during biodegradation in porous media: oxygen transfer, contaminant transfer and microbial consumption.

Oxygen transfer from the gas phase to the liquid phase can be described by the following differential equation:

$$\frac{dC}{dt} = K_L a(C^* - C) \qquad (8)$$

where C is the oxygen concentration in the liquid phase, C^* is the oxygen solubility at the given temperature, and $K_L a$ is the oxygen transfer coefficient. This coefficient is not easy to identify in a soil and depends on the degree of saturation and on the air flow rate. Zhang (Zhang, 1994) proposed the following formulation for a specific soil:

$$K_L a = c_0 + c_1\left(\frac{\theta_w}{\varepsilon}\right) + c_2\left(\frac{\theta_w}{\varepsilon}\right)^2 + c_3\left(\frac{\theta_w}{\varepsilon}\right)^3 + c_4\left(\frac{\theta_w}{\varepsilon}\right)^4 \qquad (9)$$

where ε is the porosity of the soil and c_0, c_1, c_2, c_3 and c_4 are constants.

Contaminant transfer from soil pores to the aqueous phase is a very complex phenomenon incorporating a few compartments. With the present available analytical techniques, it is impossible to quantify the substrate concentration in each compartment, a simplified approach consists on using a global substrate transfer coefficient:

$$\frac{dS}{dt} = K_S a(S^* - S) \qquad (10)$$

where S is the substrate concentration, S* is the substrate solubility at the given temperature, and $K_S a$ is the substrate transfer coefficient. Since this coefficient integrates desorption of the contaminant from the soil pores, its identification is very complicated. As for the oxygen transfer coefficient, the substrate transfer coefficient depends on the degree of saturation of the soil.

The Monod equation is commonly used to model the kinetics of biodegradation and associated biomass growth. Considering the endogenous decay rate, biomass growth and substrate depletion verify the following differential equations:

$$\frac{dS}{dt} = -Y_{S/X} \mu_m \frac{S}{K_S + S} X \qquad (11)$$

$$\frac{dX}{dt} = \mu_m \frac{S}{K_S + S} X - \mu_d X \qquad (12)$$

where S and X are respectively the substrate and biomass concentration, $Y_{S/X}$ is the substrate yield coefficient, μ_m is the maximum substrate utilization rate, K_S is the half-saturation coefficient and μ_d is the endogenous decay rate.

MATERIALS AND METHODS

Hydrodynamics Tests. Soil used for the study was a natural loamy sand (6 % clay, 7 % silt and 87 % sand). A 1.5 m high column was equipped with an irrigation and aeration system capable of providing co-current and counter-current water and air flow (Figure 1). TDR probes for water content measurement and tensio-meters for water and air pressure measurement were placed at six different depths. A data acquisition system was installed to collect measurement at every minute. A basic control system was implemented to ensure a constant water content at the top of the column. A mass flow controller was used to maintain a constant air flow through the soil.

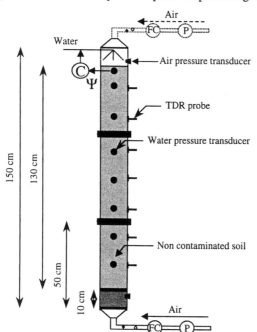

FIGURE 1. Schematic representation of the hydrodynamic column.

Tests with the 20 cm high column were performed to build the soil-water retention curve. The column, equipped with a tensiometer, was filled with soil whose humidity was known.

Biokinetic Tests. Respirometric tests were performed in triplicate in 250 mL biometric flasks with three different water contents : 20 %, 50 % and 80 % of the water holding capacity (WHC). Soil was contaminated with hexadecane at a level of 10,000 mg/kg (Figure 2). This test allowed for the assessment of carbon dioxide production, contaminant consumption (Soxtec extraction + GC/FID analysis) and biomass evolution (MPN, DNA quantification and epifluorescence microscopy). Because the results of DNA quantification and epifluorescence microscopy are not yet interpreted, they will not be discussed here. For each water content, other duplicates were performed with the injection of a radioactive marker ($^{14}C_{16}H_{34}$).

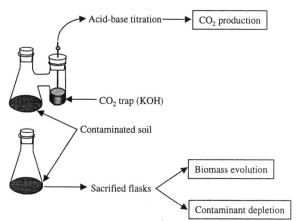

FIGURE 2. Schematic representation of the biokinetic tests.

RESULTS AND DISCUSSION

Hydrodynamic Tests. The soil-water retention curve obtained experimentally was used to identify parameters of the Van Genuchten equations. A least-square method based on the Marquard-Levenberg approach was used and lead to the following result with a regression coefficient of 0.9983: $\alpha = 0.2117$, $m = 1.3994$ and $n = 1-1/m$.

Three tests were performed on the hydrodynamic column corresponding to three operating conditions. The first test consisted in water infiltration without forced air convection. The second and third tests consisted in water percolation with respectively co-current and counter-current forced air convection. At the beginning of the tests, the soil water content was equal to 0,10 g/g dry soil, which corresponded to a soil suction of –250 cm of water. The air flow rate was fixed at 5 L/mn and the water content was maintained at 0.26 g/g dry soil at the top of the column. This water content corresponded to a soil suction of –15 cm of water. Results are presented in figure 3.

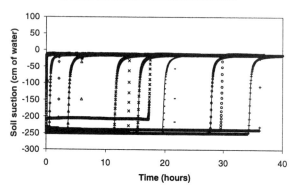
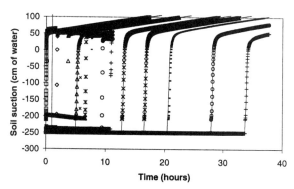
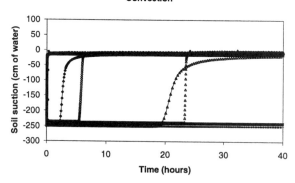

Distance from top surface : +118 cm, ○ 98.5cm, - 73 cm, ✳ 60.5 cm, × 48 cm, △ 22.5 cm, ◇ 10 cm, □ 2.5 cm

FIGURE 3. Results of hydrodynamic tests.

Air convection has a significant impact on water percolation. The co-current air flow increased water infiltration, and the counter-current air flow decreased it dramatically. On the one hand, the simulation curves, represented by solid lines, indicate that the hydrodynamic model based on Darcy's law and the Van Genuchten relations correctly predicts the water infiltration without air convection but introduces more dispersion than the actual percolation. The saturated hydraulic conductivity was adjusted to $7.2.10^{-2}$ cm/s, which is in accordance with the values commonly encountered for loamy sands. On the other hand, with air convection, the model doesn't fit accurately the data even if it predicts qualitatively the influence of air convection on water infiltration. This discrepancy between simulation and experimental data is presently under investigation but it *a priori* seems that predictive models of the unsaturated conductivities are not valid in presence of air convection.

Biokinetic Tests. Figure 4 shows the evolution with time of hexadecane mineralization obtained from the biometric flasks containing radioactivity for the three water contents. After a three-day lag phase, the higher the water content, the smaller the mineralization rate obtained. After about fifty days, the mineralization rate seems to reach a constant value, which is a little higher at 20 % of WHC than for the other two water contents. These results can be explained by the fact that, in the first fifty days, mineralization is controlled by the oxygen transfer rate. After this period, mineralization is probably controlled by the contaminant transfer rate and by the endogenous decay rate. The biomass and hexadecane evolution can enhance this interpretation. The MPN method gives the most interesting information since it allows to both quantify the heterotrophic and the hydrocarbonoclastic microbial populations. The total heterotrophic population remained constant (5.10^7 MPN index/g dry soil), while the hydrocarbon degraders became dominant after only about fifteen days, increasing from 10^5 to 5.10^7 MPN index/g dry soil. This is in accordance with the fact that the exponential phase is mainly controlled by the oxygen transfer rate.

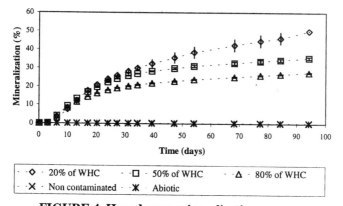

FIGURE 4. Hexadecane mineralization curves.

However the low accuracy of the MPN method doesn't allow to distinguish between the evolutions of the microbial populations for the three water contents. Furthermore, high soil heterogeneity and analytical problems prevent us from discussing hexadecane depletion. Present work is presently being undertaken to cope with the analytical problems concerning the accuracy of the MPN method and the hexadecane monitoring. This information is required to calibrate and validate the proposed model of biodegradation in unsaturated soils.

CONCLUSION

Hydrodynamic tests revealed that air flow has a significant impact on the water infiltration. The Darcy's law-based model correctly described water percolation without air convection. In presence of air flow, the model is presently under validation. The biokinetic tests indicate that water content influences hexadecane mineralization especially within the early weeks during which oxygen is probably the limiting factor. After this period, the mineralization seems to be controlled by the hexadecane transfer rate and by the endogenous decay rate. Future investigations will deal with the accurate monitoring of the microbial population and the hexadecane concentration in unsaturated soils. Biodynamic tests in a column are also being performed to assess dynamic characteristics such as the oxygen transfer coefficient, which is probably closely dependent on the air flow rate.

ACKNOWLEDGMENTS

The authors acknowledge the financial support of the Chair partners: Alcan, Bell Canada, Canadian Pacific Railway, Cambior, Centre d'expertise en analyse environnementale du Québec, Hydro-Québec, Ministère des Affaires Municipales et de la Métropole, Natural Sciences and Engineering Research Council (NSERC), Petro-Canada, Solvay, Total Fina Elf, and Ville de Montréal

REFERENCES

Bear, J. 1972. "Dynamics of Fluids in Porous Media." American Elsevier Publications, New York, NY.

Cioffi, J. C., W. R. Mahaffey and T. M. Whitlock. 1991. "Successful Solid-phase Bioremediation of Petroleum-contaminated Soil." *Remediation*. 1:373.

Dibble, J. T. and R. Bartha. 1979. "Effect of Environmental Parameters on the Biodegradation of Oil Sludge." *Applied and Environmental Microbiology*. 37(4):729-739.

Fredlund, D. G. and H. Rahardjo. 1993. "Soil Mechanics for Unsaturated Soils." Wiley-Interscience Publication, New York, NY.

Haverkamp, R., M. Vauclin, J. Touma, P. J. Wierenga and G. Vachaud. 1977. "A Comparison of Numerical Simulation Models for One-Dimensional Infiltration." *Soil Sci. Soc. Am. J.. 41*:285-294.

McDougall, J.R. and I.C. Pyrah. 1998. "Simulating Transient Infiltration in Unsaturated Soils." *Can. Geotech. J.. 35*:1093-1100.

Sims, J. L., R. C. Sims and J. E. Matthews. 1990. "Approach to Bioremediation of Contaminated Soil." *Hazardous Waste and Hazardous Materials.* 7(2):117-149.

Van Genuchten, M. Th. 1980. "A Closed-form Equation for Predicting the Hydraulic Conductivity of Unsaturated Soils." *Soil Sci. Soc. Am. J..* 44:892-898.

Zhang, Y. 1994. "Oxygen Transfer in Soil Matrices." Ph.D Thesis, Lamar University, Beaumont, TX.

COMPARISON OF EXPERIMENTAL AND SIMULTAED RESULTS OF A ROCK-BED FILTRATION MODEL

Ranjna Jindal (Suranaree University of Technology, Nakhon Ratchasima,Thailand)
Shigeo Fujii (Kyoto University, Kyoto, Japan)

ABSTRACT: A mathematical model of rock-bed filtration process in an on-site water treatment system was formulated and a numerical computation scheme was developed. Subsequently, pilot-plant experiments were conducted along a canal in Bangkok to evaluate the performance of the rock-bed filtration system under different operating conditions. Results of the pilot-plant experiments were used to verify the developed model. The simulation results for the effluent water quality and the sediment heights indicated that the developed model described the rock-bed filtration process adequately.

INTRODUCTION

A mathematical model of a rock-bed filtration process was formulated and a numerical computation scheme was developed. Subsequently, selected physical and biological processes occurring in natural water purification were incorporated into the developed model (Jindal and Fujii, 1998).

Pilot-plant experiments were conducted for on-site water treatment along a canal in Bangkok to evaluate the performance of a rock-bed filtration system under different operating conditions by changing rock size, hydraulic retention time (HRT), and aeration. Two reactor units (R1 and R2) of rectangular cross-sections (5.04m long, 0.5m wide, and 0.75m deep) were constructed from PVC sheets (10mm thickness). There were four sampling locations at equal distance along the length of each reactor in addition to influent and effluent sampling ports. Three different rock sizes, represented by large, medium, and small with equivalent diameters in approximate ranges of 8-14 cm, 12-18 cm, and 16-22 cm, respectively were used as filter media. Polluted canal water was used as influent in steady state horizontal flow in the reactors through a head tank unit (1 x 1 x 1 m). Aeration was provided through a difuser network of 1.25 cm PVC tubes installed at the bottom of each reactors. Six selected water quality indices and sediment deposits were analyzed over a period of more than one year. A total of 10 experiments were carried out in 5 test periods, each with 2 simultaneous runs, represented by R11, R21, ..., R15, and R25, respectively (Jindal and Fujii, 1999).

The main objective of this research was to evaluate the validity of a developed model by testing it with the experimental data of the pilot-plant investigations.

MODEL CALIBRATION

General Equation of the Model. The general equation representing the change in concentration of pollutant with time and distance along the direction of flow based on the mass balance can be expressed as follows:

$$\frac{\partial C}{\partial t} + U \frac{\partial C}{\partial x} = f(C) \qquad (1)$$

where:
C = instantaneous concentration at a distance X in the direction of flow, g/m^3
U = flow velocity, m/d
f(C) = function representing concentration of pollutants
t = time, d
X = distance along the reactor length, m

Eight physical and biological processes were considered in the development of a comprehensive mathematical model for the rock-bed filtration system. These processes included the reaeration, sedimentation of organic and inert particulate matter, oxic decomposition of dissolved organic matter, nitrification, and oxic, anoxic, and anaerobic decomposition of organics in sediments (Jindal and Fujii, 1998).

Kinetic and Stoichiometric Coefficients. The developed mathematical model of rock-bed filtration process was calibrated with two experimental runs by comparing the observed and simulated values of six effluent quality variables, particulate organic matter (POM), inorganic suspended solids (ISS), dissolved oxygen (DO), dissolved organic matter (DOM), and ammonium and nitrate nitrogen (NH_4^+-N, and NO_3^--N) represented by $C_1 \ldots C_6$. Observed values of POM and ISS were calculated from the experimental data of VSS and SS by using the conversion equations (Jindal and Fujii, 1998). The first two simultaneous experimental runs in two reactor units R1 and R2, represented by R11 and R21, respectively, were chosen for calibration purposes. Results of the remaining runs were used for model verification. The operating conditions of the three runs (R11, R21, and R13) are shown in Table 1.

All kinetic coefficients (i.e. half saturation constants) and stoichiometeric coefficients were either adopted from literature or determined from the observed experimental data, and are given in Table 2. In case of reaeration, while K_L represents the usual surface to water aeration coefficient, an additional gas mass transfer coefficient K_{L2} was incorporated in the process rate equation for our study. This represented the aeration from the reactor bottom.

Agreement between the six simulated and experimental effluent water quality variables for the first run (R11) is shown in Figure 1. It can be seen that the agreement for NH_4^+-N and NO_3^--N was in general not as good as for the other variables. This can be attributed to the fact that both nitrification and denitrification reactions are too sensitive to the availability and presence (or absence) of DO, and the DO levels along the reactor lengths fluctuated substantially during the experimental runs.

TABLE 1. Average operating conditions of runs R11, R21, and R13

Parameter					Run Number		
					R11	R21	R13
Influent flow rate (m^3/d)					5.76	5.76	10.8
HRT (h)					6	6	3
Initial Porosity (%)					64.12	63.4	51.9
Final Porosity (%)					55.9	53.7	50.1
Temperature (°C)					28.5	28.5	30.5
COD Loading Rate	=	Flow rate × T-COD	(kg/d)		0.38	0.38	0.86
Influent							
SS (mg/L)					63.1	65.3	82.6
VSS (mg/L)					26.6	27.0	26.7
T-COD (mg/L)					66.70	66.70	79.60
C_1	=	POM	=	1.43 × VSS (mg/L)	38.04	38.61	38.18
C_2	=	TSS	=	SS - VSS (mg/L)	36.5	38.3	55.9
C_3	=	DO		(mg/L)	1.8	1.7	3.4
C_4	=	S-COD		(mg/L)	21.4	21.4	31.0
C_5	=	NH_4^+-N		(mg/L)	5.5	5.5	11.3
C_6	=	NO_3^--N		(mg/L)	0.55	0.55	0.47

MODEL VERIFICATION

Effluent Water Quality. After model calibration and obtaining the values for all the kinetic and stoichiometric coefficients (as shown in Table 2), the model was verified against the remaining experimental runs (R12, R22, R13, R23, R14, R24, R15, and R25). Figures 2 shows the comparison of simulated and observed values of six effluent water quality parameters for run R13. For most experimental runs, experimental values of effluent SS, VSS, S-COD, NH_4^+-N, and NO_3^--N compared well with the simulated ones. However, the agreement between the observed and simulated DO was not very good. This could be attributed to the fluctuating DO levels along the reactor channels due to the difficulties in controlling the experimental conditions precisely.

Height of Sediment Deposited at the Reactor Bottom. Sedimentation is one of the major pollutant removal processes in water self-purification which affects the flow velocity due to sediment being deposited at the bottom of reactor. The change in the suspended solids concentration (ΔC) was related to the change in

TABLE 2. Values of kinetic and stoichiometric coefficients

Parameter	Value	Units	Value reported (Ref.)
d_1	5.4	d^{-1}	7.2 (1)
d_2	9.2	d^{-1}	
α_1	1.4×10^{-6}	m^3/g	Based on typical density value of sand (Organic SS)
α_2	3.85×10^{-7}	m^3/g	Based on typical density value of sand (inorganic SS)
r_1	2.4	d^{-1}	2.4 (1)
r_2	60	$g/m^2 d$	
r_3	30	$g\,N/m^3 d$	
c	0.02		
k_{34}	0.5	$g\,O_2/m^3$	0.5 (1)
k_{35}	1.0	$g\,O_2/m^3$	0.5 (1)
k_3	1.3	$g\,O_2/m^3$	1.3 (1), (2)
k_5	1.4	$g\,N/m^3$	1.4 (2)
k_6	0.5	$g\,N/m^2$	1.0 (1)
k_s	700	g/m^2	71.4 (1)
k_L	1.68	m/d	1.6 (3)
k_{L2}	8.64	m/d	8.64 (3)
c_s	8.99	g/m^3	8.99 (2)
w_1	0.124	$g\,N/g\,O_2$	Observed in experiments
w_2	0.02	$g\,N/g\,O_2$	Observed in experiments
w_3	0.7	$g\,SS/g\,O_2$	Based on emperical Formula
w_4	0.7	$g\,N/g\,O_2$	0.35 (1)
w_5	4.58	$g\,O_2/g\,NH_4\text{-}N$	4.57 (1)
b	0.5		

(1) Fujii and Somiya (1990), (2) Jindal (1995), (3) Bailey and David (1986)

where:

w_1 = N/COD ratio in DOM, $gN/g\,O_2$

w_2 = N/COD ratio in sediment organics, $gN/g\,O_2$

w_3	=	conversion factor from COD to weight
w_4	=	g N/g COD consumed
w_5	=	DO demand for nitrification, g O_2/g N
r_1, r_2, r_3	=	reaction rate constants
$k_{34}, k_{35}, k_3, k_5, k_s$	=	half saturation coefficients
k_L	=	air-water gas transfer rate constant, d^{-1}
C_s	=	saturation DO concentration at 20°C, g/m^3
b, c	=	conversion factors

the height of sediment layer (ΔY) along the distance (ΔX) in a time (Δt) as follows: (Jindal and Fujii, 1998)

$$\Delta Y = \Delta C * h * \alpha \qquad (2)$$

where:
h = initial height of net available passage for water flow in reactor, m
α = specific volume of the sediments, m^3/g

Initially, the value of h was estimated by multiplying by height of reactor by the porosity of rock-bed. In subsequent simulation steps, the values of ΔY were successively subtracted from h when calculating the flow velocity U in any given time step Δt. In order to obtain the simulated value of the height of the sediment deposited at the reactor bottoms during any run, the amount (in kg) and height of the sediment simulated during the previous run were used as initial values for the current run. For the first experimental run, the initial values of sediment amount and height were taken as zero. Thus the output of each simulated run produced cumulative values of the amount and height of the sediment until the end of that run. The simulated sediment layer heights were compared with the observed ones measured at the end of each run. However, in order to compare the observed and simulated sediment heights, one important factor needs to be considered. The simulated heights were based on the dry sediment mass deposits, whereas the measured ones were the heights of the wet sediment mass deposits. Therefore, conversion of the simulated heights from dry basis to wet basis was done using the following equation.

$$Y_{wet} = (Y_{dry} * D_{dry})/(D_{wet} * S) \qquad (3)$$

where, Y_{wet} = wet basis sediment height
Y_{dry} = dry basis sediment height
D_{wet} = density of the wet mass of the sediment
D_{dry} = density of the dry mass of the sediment
S = solid content in sediment

D_{wet} and S were determined experimentally on the sediment samples. Taking D_{dry} as the typical value of the density of sand i.e., 2.6 g cm^{-3}, above equation could be used

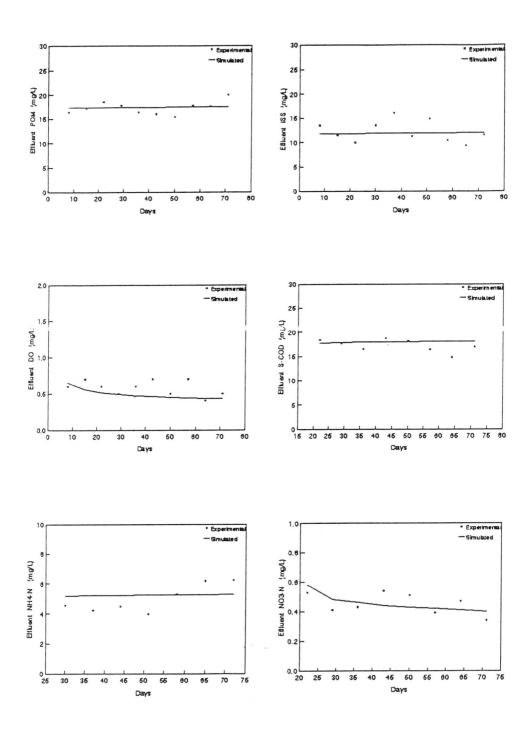

FIGURE 1. Model Calibration with Run R11.

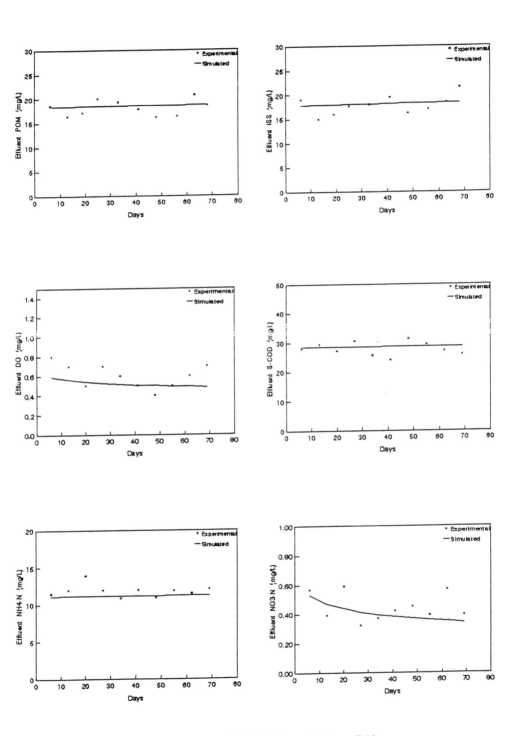

FIGURE 2. Model Verification with Run R13.

to convert the dry simulated sediment heights to the equivalent simulated wet height values. These values then could be compared with the observed values of the sediment heights measured at the end of each experimental run as shown in Figure 3 which shows a reasonably good agreement between the observed and simulated sediment heights for the runs (Run R11 – R23).

In view of the simulation results in terms of the effluent water quality indices as well as the sediment heights, it could be concluded that the developed mathematical model described the rock-bed filtration process adequately except for effluent nitrogen concentrations in runs R14, R24, and R15, i.e. when simultaneous nitrification occurred due to additional aeration at mid way along the reactor lengths.

CONCLUSIONS

The simulation results in terms of the effluent water quality indices as well as the sediment heights indicated that the developed mathematical model described the rock-bed filtration process adequately. Thus, it can be concluded that a mathematical model in the form of a set of differential equations (Jindal and Fujii, 1998) could describe the rock-bed filtration method adequately with some limitations regarding the nitrification and denitrification process in biofilm and sediments.

REFERENCES

Bailey, J.E. and F.O. David. 1986. *Biochemical Engineering Fundamentals*. International Edition, McGraw Hill Co.

Fujii S. and I. Somiya. 1990. "On a renovation strategy toward advanced treatment by a large-scale wastewater treatment plant simulation model," Preprint of the poster paper presented in the 17th IAWPRC Conference, Kyoto, Japan, pp. 803-806.

Jindal, R. and S. Fujii. 1998. "Modelling of Rock-bed Filtration Process," *Environmental Technology* 19: 273-281.

Jindal, R. and S. Fujii. 1999. "Pilot-plant Experiments on Rock-bed Filtration for improving canal water quality," *Environmental Technology* 20: 343-354.

Jindal, R. 1995. "Development and modelling of a water treatment system using the rock-bed filtration method," Ph.D.Thesis, Asian Institute of Technology, Bangkok, Thailand.

Somiya, I. 1984. *Report on purification system for comprehensive development in Inagawa basin.* (in Japanese). Purification System Committee, Kinki Regional Construction Bureau, Ministry of Construction, Japan.

Modeling

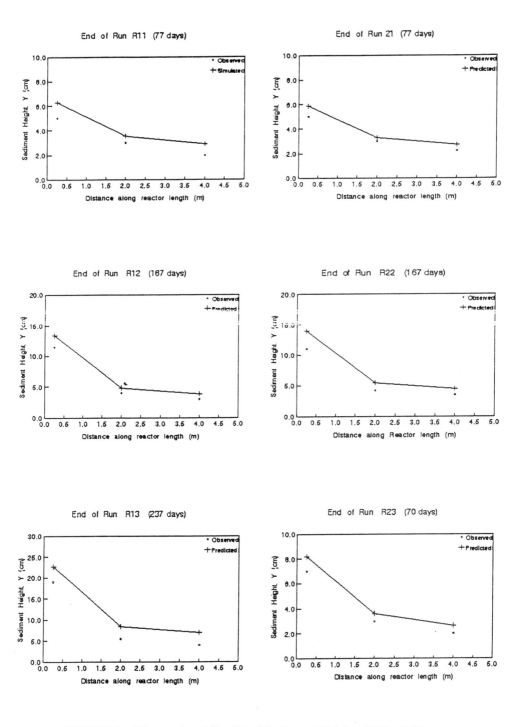

FIGURE 3. Observed and Predicted Sediment Heights (R11 - R23).

2001 AUTHOR INDEX

This index contains names, affiliations, and volume/page citations for all authors who contributed to the ten-volume proceedings of the Sixth International In Situ and On-Site Bioremediation Symposium (San Diego, California, June 4-7, 2001). Ordering information is provided on the back cover of this book. The citations reference the ten volumes as follows:

6(1): Magar, V.S., J.T. Gibbs, K.T. O'Reilly, M.R. Hyman, and A. Leeson (Eds.), *Bioremediation of MTBE, Alcohols, and Ethers*. Battelle Press, Columbus, OH, 2001. 249 pp.
6(2): Leeson, A., M.E. Kelley, H.S. Rifai, and V.S. Magar (Eds.), *Natural Attenuation of Environmental Contaminants*. Battelle Press, Columbus, OH, 2001. 307 pp.
6(3): Magar, V.S., G. Johnson, S.K. Ong, and A. Leeson (Eds.), *Bioremediation of Energetics, Phenolics, and Polycyclic Aromatic Hydrocarbons*. Battelle Press, Columbus, OH, 2001. 313 pp.
6(4): Magar, V.S., T.M. Vogel, C.M. Aelion, and A. Leeson (Eds.), *Innovative Methods in Support of Bioremediation*. Battelle Press, Columbus, OH, 2001. 197 pp.
6(5): Leeson, A., E.A. Foote, M.K. Banks, and V.S. Magar (Eds.), *Phytoremediation, Wetlands, and Sediments*. Battelle Press, Columbus, OH, 2001. 383 pp.
6(6): Magar, V.S., F.M. von Fahnestock, and A. Leeson (Eds.), *Ex Situ Biological Treatment Technologies*. Battelle Press, Columbus, OH, 2001. 423 pp.
6(7): Magar, V.S., D.E. Fennell, J.J. Morse, B.C. Alleman, and A. Leeson (Eds.), *Anaerobic Degradation of Chlorinated Solvents*. Battelle Press, Columbus, OH, 2001. 387 pp.
6(8): Leeson, A., B.C. Alleman, P.J. Alvarez, and V.S. Magar (Eds.), *Bioaugmentation, Biobarriers, and Biogeochemistry*. Battelle Press, Columbus, OH, 2001. 255 pp.
6(9): Leeson, A., B.M. Peyton, J.L. Means, and V.S. Magar (Eds.), *Bioremediation of Inorganic Compounds*. Battelle Press, Columbus, OH, 2001. 377 pp.
6(10): Leeson, A., P.C. Johnson, R.E. Hinchee, L. Semprini, and V.S. Magar (Eds.), *In Situ Aeration and Aerobic Remediation*. Battelle Press, Columbus, OH, 2001. 391 pp.

Aagaard, Per (University of Oslo/NORWAY) 6(2):181
Aarnink, Pedro J.P. (Tauw BV/THE NETHERLANDS) 6(10):253
Abbott, James E. (Battelle/USA) 6(5):231, 237
Accashian, John V. (Camp Dresser & McKee, Inc./USA) 6(7):133
Adams, Daniel J. (Camp Dresser & McKee, Inc./USA) 6(8):53

Adams, Jack (Applied Biosciences Corporation/USA) 6(9):331
Adriaens, Peter (University of Michigan/USA) 6(8):19, 193
Adrian, Neal R. (U.S. Army Corps of Engineers/USA) 6(6):133
Agrawal, Abinash (Wright State University/USA) 6(5):95
Aiken, Brian S. (Parsons Engineering Science/USA) 6(2): 65, 189

Aitchison, Eric (Ecolotree, Inc./USA) 6(5):121
Al-Awadhi, Nader (Kuwait Institute for Scientific Research/KUWAIT) 6(6):249
Alblas, B. (Logisticon Water Treatment/THE NETHERLANDS) 6(8):11
Albores, A. (CINVESTAV-IPN/MEXICO) 6(6):219
Al-Daher, Reyad (Kuwait Institute for Scientific Research/KUWAIT) 6(6):249
Al-Fayyomi, Ihsan A. (Metcalf & Eddy, Inc./USA) 6(7):173
Al-Hakak, A. (McGill University/CANADA) 6(9):139
Allen, Harry L. (U.S. EPA/USA) 6(3):259
Allen, Jeffrey (University of Cincinnati/USA) 6(9):9
Allen, Mark H. (Dames & Moore/USA) 6(10):95
Allende, J.L. (Universidad Complutense/SPAIN) 6(4):29
Alonso, R. (Universidad Politecnica/SPAIN) 6(6):377
Alphenaar, Arne (TAUW bv/THE NETHERLANDS) 6(7):297
Alvarez, Pedro J. J. (University of Iowa/USA) 6(1):195; 6(3):1; 6(8):147, 175
Alvestad, Kimberly R. (Earth Tech/USA) 6(3):17
Ambert, Jack (Battelle Europe/SWITZERLAND) 6(6):241
Amezcua-Vega, Claudia (CINVESTAV-IPN/MEXICO) 6(3):243
Amy, Penny (University of Nevada Las Vegas/USA) 6(9):257
Andersen, Peter F. (GeoTrans, Inc./USA) 6(10):163
Anderson, Bruce (Plan Real AG/AUSTRALIA) 6(2):223
Anderson, Jack W. (RMT, Inc./USA) 6(10):201
Anderson, Todd (Texas Tech University/USA) 6(9):273
Andreotti, Giorgio (ENI Sop.A.) 6(5):41

Andretta, Massimo (Centro Ricerche Ambientali Montecatini/ITALY) 6(4):131
Andrews, Eric (Environmental Management, Inc./USA) 6(10):23
Andrews, John (SHN Consulting Engineers & Geologists, Inc./USA) 6(3):83
Archibald, Brent B. (Exxon Mobil Environmental Remediation/USA) 6(8):87
Archibold, Errol (Spelman College/USA) 6(9):53
Aresta, Michele (Universita di Catania/ITALY) 6(3):149
Arias, Marianela (PDVSA Intevep/VENEZUELA) 6(6):257
Atagana, Harrison I. (Mangosuthu Technikon/REP OF SOUTH AFRICA) 6(6):101
Atta, Amena (U.S. Air Force/USA) 6(2):73
Ausma, Sandra (University of Guelph/CANADA) 6(6):185
Autenrieth, Robin L. (Texas A&M University/USA) 6(5): 17, 25
Aziz, Carol E. (Groundwater Services, Inc./USA) 6(7):19; 6(8):73
Azizian, Mohammad (Oregon State University/USA) 6(10): 145, 155

Babel, Wolfgang (UFZ Center for Environmental Research/GERMANY) 6(4):81
Bae, Bumhan (Kyungwon University/REPUBLIC OF KOREA) 6(6):51
Baek, Seung S. (Kyonggi University/REPUBLIC OF KOREA) 6(1):161
Bagchi, Rajesh (University of Cincinnati/USA) 6(5):243, 253, 261
Baiden, Laurin (Clemson University/USA) 6(7):109
Bakker, C. (IWACO/THE NETHERLANDS) 6(7):141
Balasoiu, Cristina (École Polytechnique de Montreal/CANADA) 6(9):129
Balba, M. Talaat (Conestoga-Rovers & Associates/USA) 6(1):99; 6(6):249; 6(10):131

Banerjee, Pinaki (Harza Engineering Company, Inc./USA) 6(7):157
Bankston, Jamie L. (Camp Dresser and McKee Inc./USA) 6(5):33
Barbé, Pascal (Centre National de Recherche sur les Sites et Sols Pollués/FRANCE) 6(2):129
Barcelona, Michael J. (University of Michigan/USA) 6(8):19, 193
Barczewski, Baldur (Universitat Stuttgart/GERMANY) 6(2):137
Barker, James F. (University of Waterloo/CANADA) 6(8):95
Barnes, Paul W. (Earth Tech, Inc./USA) 6(3): 17, 25
Basel, Michael D. (Montgomery Watson Harza/USA) 6(10):41
Baskunov, Boris B. (Russian Academy of Sciences/RUSSIA) 6(3):75
Bastiaens, Leen (VITO/BELGIUM) 6(4):35; 6(9):87
Batista, Jacimaria (University of Nevada Las Vegas/USA) 6(9): 257, 265
Bautista-Margulis, Raul G. (Centro de Investigacion en Materiales Avanzados/MEXICO) 6(6):361
Becker, Paul W. (Exxon Mobil Refining & Supply/USA) 6(8):87
Beckett, Ronald (Monash University/AUSTRALIA) 6(4):1
Beckwith, Walt (Solutions Industrial & Environmental Services/USA) 6(7):249
Beguin, Pierre (Institut Pasteur/FRANCE) 6(1):153
Behera, N. (Sambalpur University/INDIA) 6(9):173
Bell, Nigel (Imperial College London/UK) 6(10):123
Bell, Mike (Coats North America/USA) 6(7):213
Beller, Harry R. (Lawrence Livermore National Laboratory/USA) 6(1):195
Belloso, Claudio (Facultad Catolica de Quimica e Ingenieria/ARGENTINA) 6(6): 235, 303
Benner, S. G. (Stanford University/USA) 6(9):71
Bensch, Jeffrey C. (GeoTrans, Inc/USA) 6(7):221

Béron, Patrick (Université du Québec à Montréal/CANADA) 6(3):165
Berry, Duane F. (Virginia Polytechnic Institute & State University/USA) 6(2):105
Betts, W. Bernard (Cell Analysis Ltd./UK) 6(6):27
Billings, Bradford G. (Brad) (Billings & Associates, Inc./USA) 6(1):115
Bingler, Linda (Battelle Sequim/USA) 6(5):231, 237
Birkle, M. (Fraunhofer Institute/GERMANY) 6(2):137
Bitter, Paul (URS Corporation./USA) 6(2):261
Bittoni, A. (EniTecnologie/ITALY) 6(6):173
Bjerg, Poul L (Technical University of Denmark/DENMARK) 6(2):11
Blanchet, Denis (Institut Français du Pétrole/FRANCE) 6(3):227
Bleckmann, Charles A. (Air Force Institute of Technology/USA) 6(2):173
Blokzijl, R. (DHV Environment and Infrastructure/THE NETHERLANDS) 6(8):11
Blowes, David (University of Waterloo/CANADA) 6(9):71
Bluestone, Simon (Montgomery Watson/ITALY) 6(10):41
Boben, Carolyn (Williams/USA) 6(1):175
Böckle, Karin (Technologiezentrum Wasser/GERMANY) 6(8):105
Boender, H. (Logisticon Water Treatment/THE NETHERLANDS) 6(8):11
Böhler, Anja (BioPlanta GmbH/GERMANY) 6(3):67
Bonner, James S. (Texas A&M University/USA) 6(5):17, 25
Bononi, Vera Lucia Ramos (Instituto de Botânica/BRAZIL) 6(3):99
Bonsack, Laurence T. (Aerojet/USA) 6(9):297
Borazjani, Abdolhamid (Mississippi State University/USA) 6(5):329; 6(6):279

Borden, Robert C. (Solutions Industrial & Environmental Services/USA) 6(7):249
Bornholm, Jon (U.S. EPA/USA) 6(6):81
Bosco, Francesca (Politecnico di Torino/ITALY) 6(3):211
Bosma, Tom N.P. (TNO Environment/THE NETHERLANDS) 6(7):61
Bourquin, Al W. (Camp Dresser & McKee Inc./USA) 6(5):33; 6(6):81; 6(7):133,
Bouwer, Edward J. (Johns Hopkins University/USA) 6(2):19
Bowman, Robert S. (New Mexico Institute of Mining & Technology/USA) 6(8):131
Boyd, Sian (CEFAS Laboratory/UK) 6(10):337
Boyd-Kaygi, Patricia (Harding ESE/USA) 6(10):231
Boyle, Susan L. (Haley & Aldrich, Inc./USA) 6(7):27, 281
Brady, Warren D. (IT Corporation/USA) 6(9):215
Breedveld, Gijs (University of Oslo/NORWAY) 6(2):181
Bregante, M. (Istituto di Cibernetica e Biofisica/ITALY) 6(5):157
Brenner, Richard C. (U.S. EPA/USA) 6(5):231, 237
Breteler, Hans (Oostwaardhoeve Co./THE NETHERLANDS) 6(6):59
Bricka, Mark R. (U.S. Army Corps of Engineers/USA) 6(9):241
Brickell, James L. (Earth Tech, Inc./USA) 6(10):65
Brigmon, Robin L. (Westinghouse Savannah River Co/USA) 6(7):109
Britto, Ronnie (EnSafe, Inc./USA) 6(9):315
Brossmer, Christoph (Degussa Corporation/USA) 6(10):73
Brown, Bill (Dunham Environmental Services/USA) 6(6):35
Brown, Kandi L. (IT Corporation/USA) 6(1):51
Brown, Richard A. (ERM, Inc./USA) 6(7):45, 213
Brown, Stephen (Queen's University/CANADA) 6(2):121

Brown, Susan (National Water Research Institute/CANADA) 6(7):321, 333, 341
Brubaker, Gaylen (ThermoRetec North Carolina Corp./USA) 6(7):1
Bruce, Cristin (Arizona State University/USA) 6(8):61
Bruce, Neil C. (University of Cambridge/UK) 6(5):69
Buchanan, Gregory (Tait Environmental Management, Inc./USA) 6(10):267
Bucke, Christopher (University of Westminster/UK) 6(3):75
Bulloch, Gordon (BAE Systems Properties Ltd./UK) 6(6):119
Burckle, John (U.S. EPA/USA) 6(9):9
Burden, David S. (U.S. EPA/USA) 6(2):163
Burdick, Jeffrey S. (ARCADIS Geraghty & Mills/USA) 6(7):53
Burgos, William (The Pennsylvania State University/USA) 6(8):201
Burken, Joel G. (University of Missouri-Rolla/USA) 6(5):113, 199
Burkett, Sharon E. (ENVIRON International Corp./USA) 6(7):189
Burnell, Daniel K. (GeoTrans, Inc./USA) 6(2):163
Burns, David A. (ERM, Inc./USA) 6(7):213
Burton, Christy D. (Battelle/USA) 6(1):137; 6(10):193
Buscheck, Timothy E. (Chevron Research & Technology Co/USA) 6(1): 35, 203
Buss, James A. (RMT, Inc./USA) 6(2):97
Butler, Adrian P. (Imperial College London/UK) 6(10):123
Butler, Jenny (Battelle/USA) 6(7):13
Büyüksönmez, Fatih (San Diego State University/USA) 6(10):301

Caccavo, Frank (Whitworth College/USA) 6(8):1
Callender, James S. (Rockwell Automation/USA) 6(7):133
Calva-Calva, G. (CINVESTAV-IPN/MEXICO) 6(6):219
Camper, Anne K. (Montana State University/USA) 6(7):117

Camrud, Doug (Terracon/USA) 6(10):15
Canty, Marietta C. (MSE Technology Applications/USA) 6(9):35
Carman, Kevin R. (Louisiana State University/USA) 6(5):305
Carrera, Paolo (Ambiente S.p.A./ITALY) 6(6):227
Carson, David A. (U.S. EPA/USA) 6(2):247
Carvalho, Cristina (Clemson University/USA) 6(7):109
Case, Nichole L. (Haley & Aldrich, Inc./USA) 6(7):27, 281
Castelli, Francesco (Universita di Catania/ITALY) 6(3):149
Cha, Daniel K. (University of Delaware/USA) 6(6):149
Chaney, Rufus L. (U.S. Department of Agriculture/USA) 6(5):77
Chang, Ching-Chia (National Chung Hsing University/TAIWAN) 6(10):217
Chang, Soon-Woong (Kyonggi University/REPUBLIC OF KOREA) 6(1):161
Chang, Wook (University of Maryland/USA) 6(3):205
Chapuis, R. P. (École Polytechnique de Montréal/CANADA) 6(4):139
Charrois, Jeffrey W.A. (Komex International, Ltd./CANADA) 6(4):7
Chatham, James (BP Exploration/USA) 6(2):261
Chekol, Tesema (University of Maryland/USA) 6(5):77
Chen, Abraham S.C. (Battelle/USA) 6(10):245
Chen, Chi-Ruey (Florida International University/USA) 6(10):187
Chen, Zhu (The University of New Mexico/USA) 6(9):155
Cherry, Jonathan C. (Kennecott Utah Copper Corp/USA) 6(9):323
Child, Peter (Investigative Science Inc./CANADA) 6(2):27
Chino, Hiroyuki (Obayashi Corporation/JAPAN) 6(6):249
Chirnside, Anastasia E.M. (University of Delaware/USA) 6(6):9

Chiu, Pei C. (University of Delaware/USA) 6(6):149
Cho, Kyung-Suk (Ewha University/REPUBLIC OF KOREA) 6(6):51
Choung, Youn-kyoo (Yonsei University/REPUBLIC OF KOREA) 6(6):51
Clement, Bernard (École Polytechnique de Montréal/CANADA) 6(9):27
Clemons, Gary (CDM Federal Programs Corp./USA) 6(6):81
Cocos, Ioana A. (École Polytechnique de Montréal/CANADA) 6(9):27
Cocucci, M. (Universita' degli Studi di Milano/ITALY) 6(5):157
Coelho, Rodrigo O. (CSD-GEOLOCK/BRAZIL) 6(1):27
Collet, Berto (TAUW bv/THE NETHERLANDS) 6(10):253
Compton, Joanne C. (REACT Environmental Engineers/USA) 6(3):25
Connell, Doug (Barr Engineering Company/USA) 6(5):105
Connor, Michael A. (University of Melbourne/AUSTRALIA) 6(10):329
Cook, Jim (Beazer East, Inc./USA) 6(2):239
Cooke, Larry (NOVA Chemicals Corporation/USA) 6(4):117
Coons, Darlene (Conestoga-Rovers & Associates/USA) 6(1):99; 6(10):131
Costley, Shauna C. (University of Natal/REP OF SOUTH AFRICA) 6(9):79
Cota, Jennine L. (ARCADIS Geraghty & Miller, Inc./USA) 6(7):149
Covell, James R. (EG&G Technical Services, Inc./USA) 6(10):49
Cowan, James D. (Ensafe Inc./USA) 6(9):315
Cox, Evan E. (GeoSyntec Consultants/CANADA) 6(8):27, 6(9):297
Cox, Jennifer (Clemson University/USA) 6(7):109
Craig, Shannon (Beazer East, Inc./USA) 6(2):239
Crawford, Donald L. (University of Idaho/USA) 6(3):91; 6(9):147

Crecelius, Eric (Battelle/USA) 6(5): 231, 237
Crotwell, Terry (Solutions Industrial & Environmental Services/USA) 6(7):249
Cui, Yanshan (Chinese Academy of Sciences/CHINA) 6(9):113
Cunningham, Al B. (Montana State University/USA) 6(7):117; 6(8):1
Cunningham, Jeffrey A. (Stanford University/USA) 6(7):95
Cutright, Teresa J. (The University of Akron/USA) 6(3):235

da Silva, Marcio Luis Busi (University of Iowa/USA) 6(1):195
Daly, Daniel J. (Energy & Environmental Research Center/USA) 6(5):129
Daniel, Fabien (AEA Technology Environment/UK) 6(10):337
Daniels, Gary (GeoTrans/USA) 6(8):19
Das, K.C. (University of Georgia/USA) 6(9):289
Davel, Jan L. (University of Cincinnati/USA) 6(6):133
Davis, Gregory A. (Microbial Insights Inc./USA) 6(2):97
Davis, Jeffrey L. (U.S. Army/USA) 6(3): 43, 51
Davis, John W. (The Dow Chemical Company/USA) 6(2):89
Davis-Hoover, Wendy J. (U.S. EPA/USA) 6(2):247
De'Ath, Anna M. (Cranfield University/UK) 6(6):329
Dean, Sean (Camp Dresser & McKee. Inc/USA) 6(7):133
DeBacker, Dennis (Battelle/USA) 6(10):145
DeHghi, Benny (Honeywell International Inc./USA) 6(2):39;6(10):283
de Jong, Jentsje (TAUW BV/THE NETHERLANDS) 6(10):253
Del Vecchio, Michael (Envirogen, Inc./USA) 6(9):281
Delille, Daniel (CNRS/FRANCE) 6(2):57
DeLong, George (AIMTech/USA) 6(7):321, 333, 341
Demers, Gregg (ERM/USA) 6(7):45
De Mot, Rene (Catholic University of Leuven/BELGIUM) 6(4):35

Deobald, Lee A. (University of Idaho/USA) 6(9):147
Deschênes, Louise (École Polytechnique de Montréal/CANADA) 6(3):115; 6(9):129
Dey, William S. (Illinois State Geological Survey/USA) 6(9):179
Díaz-Cervantes, Dolores (CINVESTAV-IPN/MEXICO) 6(6):369
Dick, Vincent B. (Haley & Aldrich, Inc./USA) 6(7):27, 281
Diehl, Danielle (The University of New Mexico/USA) 6(9):155
Diehl, Susan V. (Mississippi State University/USA) 6(5):329
Diels, Ludo (VITO/BELGIUM) 6(9):87
DiGregorio, Salvatore (University della Calabria/ITALY) 6(4):131
Di Gregorio, Simona (Universita degli Studi di Verona/ITALY) 6(3):267
Dijkhuis, Edwin (Bioclear/THE NETHERLANDS) 6(5):289
Di Leo, Cristina (EniTecnologie/ITALY) 6(6):173
Dimitriou-Christidis, Petros (Texas A& M University) 6(5):17
Dixon, Robert (Montgomery Watson/ITALY) 6(10):41
Dobbs, Gregory M. (United Technologies Research Center/USA) 6(7):69
Doherty, Amy T. (GZA GeoEnvironmental, Inc./USA) 6(7):165
Dolan, Mark E. (Oregon State University/USA) 6(10):145, 155, 179
Dollhopf, Michael (Michigan State University/USA) 6(8):19
Dondi, Giovanni (Water & Soil Remediation S.r.l./ITALY) 6(6):179
Dong, Yiting (Chinese Academy of Sciences/CHINA) 6(9):113
Dooley, Maureen A. (Regenesis/USA) 6(7):197
Dottridge, Jane (Komex Europe Ltd./UK) 6(4):17
Dowd, John (University of Georgia/USA) 6(9):289
Doughty, Herb (U.S. Navy/USA) 6(10):1

Doze, Jacco (RIZA/THE
NETHERLANDS) 6(5):289
Dragich, Brian (California Polytechnic
State University/USA) 6(2):1
Drake, John T. (Camp Dresser & McKee
Inc./USA) 6(7):273
Dries, Victor (Flemish Public Waste
Agency/BELGIUM) 6(7):87
Du, Yan-Hung (National Chung Hsing
University/TAIWAN) 6(6):353
Dudal, Yves (École Polytechnique de
Montréal/CANADA) 6(3):115
Duffey, J. Tom (Camp Dresser & McKee
Inc./USA) 6(5):33
Duffy, Baxter E. (Inland Pollution
Services, Inc./USA) 6(7):313
Duijn, Rik (Oostwaardhoeve Co./THE
NETHERLANDS) 6(6):59
Durant, Neal D. (GeoTrans, Inc./USA)
6(2):19, 163
Durell, Gregory (Battelle Ocean
Sciences/USA) 6(5):231
Dworatzek, S. (University of
Toronto/CANADA) 6(8):27
Dwyer, Daryl F. (University of
Minnesota/USA) 6(3):219
Dzantor, E. K. (University of
Maryland/USA) 6(5):77

Ebner, R. (GMF/GERMANY) 6(2):137
Ederer, Martina (University of
Idaho/USA) 6(9):147
Edgar, Michael (Camp Dresser &
McKee Inc./USA) 6(7):133
Edwards, Elizabeth A. (University of
Toronto/CANADA) 6(8):27
Edwards, Grant C. (University of
Guelph/CANADA) 6(6):185
Eggen, Trine (Jordforsk Centre for Soil
and Environmental
Research/NORWAY) 6(6):157
Eggert, Tim (CDM Federal Programs
Corp./USA) 6(6):81
Elberson, Margaret A. (DuPont
Co./USA) 6(8):43
Elliott, Mark (Virginia Polytechnic
Institute & State University/USA)
6(5):1
Ellis, David E. (Dupont Company/USA)
6(8):43

Ellwood, Derek C. (University of
Southampton/UK) 6(9):61
Else, Terri (University of Nevada Las
Vegas/USA) 6(9):257
Elväng, Annelie M. (Stockholm
University/SWEDEN) 6(3):133
England, Kevin P. (USA) 6(5):105
Ertas, Tuba Turan (San Diego State
University/USA) 6(10):301
Escalon, Lynn (U.S. Army Corps of
Engineers/USA) 6(3):51
Esparza-Garcia, Fernando
(CINVESTAV-IPN/MEXICO)
6(6):219
Evans, Christine S. (University of
Westminster/UK) 6(3):75
Evans, Patrick J. (Camp Dresser &
McKee, Inc./USA) 6(2):113, 199;
6(8):209

Fabiani, Fabio (EniTecnologie
S.p.A./ITALY) 6(6):173
Fadullon, Frances Steinacker (CH2M
Hill/USA) 6(3):107
Fang, Min (University of
Massachusetts/USA) 6(6):73
Faris, Bart (New Mexico Environmental
Department/USA) 6(9):223
Farone, William A. (Applied Power
Concepts, Inc./USA) 6(7):103
Fathepure, Babu Z. (Oklahoma State
University/USA) 6(8):19
Faust, Charles (GeoTrans, Inc./USA)
6(2):163
Fayolle, Françoise (Institut Français du
Pétrole/FRANCE) 6(1):153
Feldhake, David (University of
Cincinnati/USA) 6(2):247
Felt, Deborah (Applied Research
Associates, Inc./USA) 6(7):125
Feng, Terry H. (Parsons Engineering
Science, Inc./USA) 6(2):39;
6(10):283
Fenwick, Caroline (Aberdeen
University/UK) 6(2):223
Fernandez, Jose M. (University of
Iowa/USA) 6(1):195
Fernández-Sanchez, J. Manuel
(CINVESTAV-IPN/MEXICO)
6(6):369

Ferrer, E. (Universidad Complutense de Madrid/SPAIN) 6(4):29
Ferrera-Cerrato, Ronald (Colegio de Postgraduados/MEXICO) 6(6):219
Fiacco, R. Joseph (Environmental Resources Management) 6(7):45
Fields, Jim (University of Georgia/USA) 6(9):289
Fields, Keith A. (Battelle/USA) 6(10):1
Fikac, Paul J. (Jacobs Engineering Group, Inc./USA) 6(6):35
Fischer, Nick M. (Aquifer Technology/USA) 6(8):157, 6(10):15
Fisher, Angela (The Pennsylvania State University/USA) 6(8):201
Fisher, Jonathan (Environment Agency/UK) 6(4):17
Fitch, Mark W. (University of Missouri-Rolla/USA) 6(5):199
Fleckenstein, Janice V. (USA) 6(6):89
Fleischmann, Paul (ZEBRA Environmental Corp./USA) 6(10):139
Fletcher, John S. (University of Oklahoma/USA) 6(5):61
Foget, Michael K. (SHN Consulting Engineers & Geologists, Inc./USA) 6(3):83
Foley, K.L. (U.S. Army Engineer Research & Development Center/USA) 6(5):9
Follner, Christina G. (University of Leipzig/GERMANY) 6(4):81
Fontenot, Martin M. (Syngenta Crop Protection, Inc./USA) 6(6):35
Foote, Eric A. (Battelle/USA) 6(1):137; 6(7):13
Ford, James (Investigative Science Inc./CANADA) 6(2):27
Forman, Sarah R. (URS Corporation/USA) 6(7):321, 333, 341
Fortman, Tim J. (Battelle Marine Sciences Laboratory/USA) 6(3):157
Francendese, Leo (U.S. EPA/USA) 6(3):259
Francis, M. McD. (NOVA Research & Technology Center/CANADA) 6(4):117; 6(5):53,
François, Alan (Institut Français du Pétrole/FRANCE) 6(1):153

Frankenberger, William T. (University of California/USA) 6(9):249
Freedman, David L. (Clemson University/USA) 6(7):109
French, Christopher E. (University of Cambridge/UK) 6(5):69
Friese, Kurt (UFZ Center for Environmental Research/GERMANY) 6(9):43
Frisbie, Andrew J. (Purdue University/USA) 6(3):125
Frisch, Sam (Envirogen Inc./USA) 6(9):281
Frömmichen, René (UFZ Centre for Environmental Research/GERMANY) 6(9):43
Fuierer, Alana M. (New Mexico Institute of Mining & Technology/USA) 6(8):131
Fujii, Kensuke (Obayashi Corporation/JAPAN) 6(10):239
Fujii, Shigeo (Kyoto University/JAPAN) 6(4):149
Furuki, Masakazu (Hyogo Prefectural Institute of Environmental Science/JAPAN) 6(5):321

Gallagher, John R. (University of North Dakota/USA) 6(5):129; 6(6):141
Gambale, Franco (Istituto di Cibernetica e Biofisica/ITALY) 6(5):157
Gambrell, Robert P. (Louisiana State University/USA) 6(5):305
Gandhi, Sumeet (University of Iowa/USA) 6(8):147
Garbi, C. (Universidad Complutense de Madrid/SPAIN) 6(4):29; 6(6):377
García-Arrazola, Roeb (CINVESTAV-IPN/MEXICO) 6(6):369
García-Barajas, Rubén Joel (ESIQIE-IPN/MEXICO) 6(6):369
Garrett, Kevin (Harding ESE/USA) 6(7):205
Garry, Erica (Spelman College/USA) 6(9):53
Gavaskar, Arun R. (Battelle/USA) 6(7):13
Gavinelli, Marco (Ambiente S.p.A./ITALY) 6(6):227
Gebhard, Michael (GeoTrans/USA) 6(8):19

Author Index

Gec, Bob (Degussa Canada Ltd./CANADA) 6(10):73
Gehre, Matthias (UFZ - Centre for Environmental Research/GERMANY) 6(4):99
Gemoets, Johan (VITO/BELGIUM) 6(4):35; 6(9):87
Gent, David B. (U.S. Army Corps of Engineers/USA) 6(9):241
Gentry, E. E. (Science Applications International Corporation/USA) 6(8):27
Georgiev, Plamen S. (University of Mining & Geology/BULGARIA) 6(9):97
Gerday, Charles (Université de Liège/BELGIUM) 6(2):57
Gerlach, Robin (Montana State University/USA) 6(8):1
Gerritse, Jan (TNO Environmental Sciences/THE NETHERLANDS) 6(2):231; 6(7):61
Gerth, André (BioPlanta GmbH/GERMANY) 6(3):67; 6(5):173
Ghosh, Upal (Stanford University/USA) 6(3):189; 6(6):89
Ghoshal, Subhasis (McGill University/CANADA) 6(9):139
Gibbs, James T. (Battelle/USA) 6(1):137
Gibello, A. (Universidad Complutense/SPAIN) 6(4):29
Giblin, Tara (University of California/USA) 6(9):249
Gilbertson, Amanda W. (University of Missouri-Rolla/USA) 6(5):199
Gillespie, Rick D. (Regenesis/USA) 6(1):107
Gillespie, Terry J. (University of Guelph/CANADA) 6(6):185
Glover, L. Anne (Aberdeen University /UK) 6(2):223
Goedbloed, Peter (Oostwaardhoeve Co./THE NETHERLANDS) 6(6):59
Golovleva, Ludmila A. (Russian Academy of Sciences/RUSSIA) 6(3):75
Goltz, Mark N. (Air Force Institute of Technology/USA) 6(2):173

Gong, Weiliang (The University of New Mexico/USA) 6(9):155
Gossett, James M. (Cornell University/USA) 6(4):125
Govind, Rakesh (University of Cincinnati/USA) 6(5):269; 6(8):35; 6(9):1, 9, 17
Gozan, Misri (Water Technology Center/GERMANY) 6(8):105
Grainger, David (IT Corporation/USA) 6(1):51; 6(2):73
Grandi, Beatrice (Water & Soil Remediation S.r.l./ITALY) 6(6):179
Granley, Brad A. (Leggette, Brashears, & Graham/USA) 6(10):259
Grant, Russell J. (University of York/UK) 6(6):27
Graves, Duane (IT Corporation/USA) 6(2):253; 6(4):109; 6(9):215
Green, Chad E. (University of California/USA) 6(10):311
Green, Donald J. (USAG Aberdeen Proving Ground/USA) 6(7):321, 333, 341
Green, Robert (Alcoa/USA) 6(6):89
Green, Roger B. (Waste Management, Inc./USA) 6(2):247; 6(6):127
Gregory, Kelvin B. (University of Iowa/USA) 6(3):1
Griswold, Jim (Construction Analysis & Management, Inc./USA) 6(1):115
Groen, Jacobus (Vrije Universiteit/THE NETHERLANDS) 6(4):91
Groenendijk, Gijsbert Jan (Hoek Loos bv/THE NETHERLANDS) 6(7):297
Grotenhuis, Tim (Wageningen Agricultural University/THE NETHERLANDS) 6(5):289
Groudev, Stoyan N. (University of Mining & Geology/BULGARIA) 6(9):97
Guarini, William J. (Envirogen, Inc./USA) 6(9):281
Guieysse, Benoît (Lund University/SWEDEN) 6(3):181
Guiot, Serge R. (Biotechnology Research Institute/CANADA) 6(3):165
Gunsch, Claudia (Clemson University/USA) 6(7):109
Gurol, Mirat (San Diego State University/USA) 6(10):301

Ha, Jeonghyub (University of Maryland/USA) 6(10):57
Haak, Daniel (RMT, Inc./USA) 6(10):201
Haas, Patrick E. (Mitretek Systems/USA) 6(7):19, 241, 249; 6(8):73
Haasnoot, C. (Logisticon Water Treatment/THE NETHERLANDS) 6(8):11
Habe, Hiroshi (The University of Tokyo/JAPAN) 6(4):51; 6(6):111
Haeseler, Frank (Institut Français du Pétrole/FRANCE) 6(3):227
Haff, James (Meritor Automotive, Inc./USA) 6(7):173
Haines, John R. (U.S. EPA/USA) 6(9):17
Håkansson, Torbjörn (Lund University/SWEDEN) 6(9):123
Halfpenny-Mitchell, Laurie (University of Guelph/CANADA) 6(6):185
Hall, Billy (Newfields, Inc./USA) 6(5):189
Hampton, Mark M. (Groundwater Services/USA) 6(8):73
Hannick, Nerissa K. (University of Cambridge/UK) 6(5):69
Hannigan, Mary (Mississippi State University) 6(5):329; 6(6):279
Hannon, LaToya (Spelman College/USA) 6(9):53
Hansen, Hans C. L. (Hedeselskabet /DENMARK) 6(2):11
Hansen, Lance D. (U.S. Army Corps of Engineers/USA) 6(3):9, 43, 51; 6(4):59; 6(6):43; 6(7):125; 6(10):115
Haraguchi, Makoo (Sumitomo Marine Research Institute/JAPAN) 6(10):345
Hardisty, Paul E. (Komex Europe, Ltd./ENGLAND) 6(4):17
Harmon, Stephen M. (U.S. EPA/USA) 6(9):17
Harms, Hauke (Swiss Federal Institute of Technology/SWITZERLAND) 6(3):251
Harmsen, Joop (Alterra, Wageningen University and Research Center/THE NETHERLANDS) 6(5):137, 279; 6(6):1, 59

Harper, Greg (TetraTech EM Inc./USA) 6(3):259
Harrington-Baker, Mary Ann (MSE, Inc./USA) 6(9):35
Harris, Benjamin Cord (Texas A&M University/USA) 6(5):17, 25
Harris, James C. (U.S. EPA/USA) 6(6):287, 295
Harris, Todd (Mason and Hanger Corporation/USA) 6(3):35
Harrison, Patton B. (American Airlines/USA) 6(1):121
Harrison, Susan T.L. (University of Cape Town/REP OF SOUTH AFRICA) 6(6):339
Hart, Barry (Monash University/AUSTRALIA) 6(4):1
Hartzell, Kristen E. (Battelle/USA) 6(1):137; 6(10):193
Harwood, Christine L. (Michael Baker Corporation/USA) 6(2):155
Hassett, David J. (Energy & Environmental Research Center/USA) 6(5):129
Hater, Gary R. (Waste Management Inc./USA) 6(2):247
Hausmann, Tom S. (Battelle Marine Sciences Laboratory/USA) 6(3):157
Hawari, Jalal (National Research Council of Canada/CANADA) 6(9):139
Hayes, Adam J. (Triple Point Engineers, Inc./USA) 6(1):183
Hayes, Dawn M. (U.S. Navy/USA) 6(3):107
Hayes, Kim F. (University of Michigan/USA) 6(8):193
Haynes, R.J. (University of Natal/REP OF SOUTH AFRICA) 6(6):101
Heaston, Mark S. (Earth Tech/USA) 6(3):17, 25
Hecox, Gary R. (University of Kansas/USA) 6(4):109
Heebink, Loreal V. (Energy & Environmental Research Center/USA) 6(5):129
Heine, Robert (EFX Systems, Inc./USA) 6(8):19
Heintz, Caryl (Texas Tech University/USA) 6(3):9

Author Index

Hendrickson, Edwin R. (DuPont Co./USA) 6(8):27, 43
Hendriks, Willem (Witteveen+Bos Consulting Engineers/THE NETHERLANDS) 6(5):289
Henkler, Rolf D. (ICI Paints/UK) 6(2):223
Henny, Cynthia (University of Maine/USA) 6(8):139
Henry, Bruce M. (Parsons Engineering Science, Inc/USA) 6(7):241
Henssen, Maurice J.C. (Bioclear Environmental Biotechnology/THE NETHERLANDS) 6(8):11
Herson, Diane S. (University of Delaware/USA) 6(6):9
Hesnawi, Rafik M. (University of Manitoba/CANADA) 6(6):165
Hetland, Melanie D. (Energy & Environmental Research Center/USA) 6(5):129
Hickey, Robert F. (EFX Systems, Inc./USA) 6(8):19
Hicks, Patrick H. (ARCADIS/USA) 6(1):107
Hiebert, Randy (MSE Technology Applications, Inc./USA) 6(8):79
Higashi, Teruo (University of Tsukuba/JAPAN) 6(9):187
Higgins, Mathew J. (Bucknell University/USA) 6(2):105
Higinbotham, James H. (ExxonMobil Environmental Remediation/USA) 6(8):87
Hines, April (Spelman College/USA) 6(9):53
Hinshalwood, Gordon (Delta Environmental Consultants, Inc./USA) 6(1):43
Hirano, Hiroyuki (The University of Tokyo/JAPAN) 6(6):111
Hirashima, Shouji (Yakult Pharmaceutical Industry/JAPAN) 6(10):345
Hirsch, Steve (Environmental Protection Agency/USA) 6(5):207
Hiwatari, Takehiko (National Institute for Environmental Studies/JAPAN) 6(5):321
Hoag, Rob (Conestoga-Rovers & Associates/USA) 6(1):99

Hoelen, Thomas P. (Stanford University/USA) 6(7):95
Hoeppel, Ronald E. (U.S. Navy/USA) 6(10):245
Hoffmann, Johannes (Hochtief Umwelt GmbH/GERMANY) 6(6):227
Hoffmann, Robert E. (Chevron Canada Resources/CANADA) 6(6):193
Höfte, Monica (Ghent University/BELGIUM) 6(5):223
Holder, Edith L. (University of Cincinnati/USA) 6(2):247
Holm, Thomas R. (Illinois State Water Survey/USA) 6(9):179
Holman, Hoi-Ying (Lawrence Berkeley National Laboratory/USA) 6(4):67
Holoman, Tracey R. Pulliam (University of Maryland/USA) 6(3):205
Hopper, Troy (URS Corporation/USA) 6(2):239
Hornett, Ryan (NOVA Chemicals Corporation/USA) 6(4):117
Hosangadi, Vitthal S. (Foster Wheeler Environmental Corp./USA) 6(9):249
Hough, Benjamin (Tetra Tech EM, Inc./USA) 6(10):293
Hozumi, Toyoharu (Oppenheimer Biotechnology/JAPAN) 6(10):345
Huang, Chin-I (National Chung Hsing University/TAIWAN) 6(10):217
Huang, Chin-Pao (University of Delaware/USA) 6(6):9, 149
Huang, Hui-Bin (DuPont Co./USA) 6(8):43
Huang, Junqi (Air Force Institute of Technology/USA) 6(2):173
Huang, Wei (University of Sheffield/UK) 6(2):207
Hubach, Cor (DHV Noord Nederland/THE NETHERLANDS) 6(8):11
Huesemann, Michael H. (Battelle/USA) 6(3):157
Hughes, Joseph B. (Rice University/USA) 6(5):85; 6(7):19
Hulsen, Kris (University of Ghent/BELGIUM) 6(5):223
Hunt, Jonathan (Clemson University/USA) 6(7):109

Hunter, William J. (U.S. Dept of Agriculture/USA) *6*(9):209, 309
Hwang, Sangchul (University of Akron/USA) *6*(3):235
Hyman, Michael R. (North Carolina State University/USA) *6*(1): 83, 145

Ibeanusi, Victor M. (Spelman College/USA) *6*(9):53
Ickes, Jennifer (Battelle/USA) *6*(5):231, 237
Ide, Kazuki (Obayashi Corporation Ltd./JAPAN) *6*(6):111; *6*(10):239
Igarashi, Tsuyoshi (Nippon Institute of Technology/JAPAN) *6*(5):321
Infante, Carmen (PDVSA Intevep/VENEZUELA) *6*(6):257
Ingram, Sherry (IT Corporation/USA) *6*(4):109
Ishikawa, Yoji (Obayashi Corporation/JAPAN) *6*(6):249; *6*(10):239

Jackson, W. Andrew (Texas Tech University/USA) *6*(5):207, 313; *6*(9):273
Jacobs, Alan K. (EnSafe, Inc./USA) *6*(9):315
Jacques, Margaret E. (Rowan University/USA) *6*(5):215
Jahan, Kauser (Rowan University/USA) *6*(5):215
James, Garth (MSE Inc./USA) *6*(8):79
Jansson, Janet K. (Södertörn University College/SWEDEN) *6*(3):133
Japenga, Jan (Alterra/THE NETHERLANDS) *6*(5):137
Jauregui, Juan (Universidad Nacional Autonoma de Mexico/MEXICO) *6*(6):17
Jensen, James N. (State University of New York at Buffalo/USA) *6*(6):89
eon, Mi-Ae (Texas Tech University/USA) *6*(9):273
Jerger, Douglas E. (IT Corporation/USA) *6*(3):35
Jernberg, Cecilia (Södertörn University College/SWEDEN) *6*(3):133
Jindal, Ranjna (Suranaree University of Technology/THAILAND) *6*(4):149

Johnson, Dimitra (Southern University at New Orleans/USA) *6*(5):151
Johnson, Glenn (University of Utah/USA) *6*(5):231
Johnson, Paul C. (Arizona State University/USA) *6*(1):11; *6*(8):61
Johnson, Richard L. (Oregon Graduate Institute/USA) *6*(10):293
Jones, Antony (Komex H_2O Science, Inc./USA) *6*(2):223; *6*(3):173; *6*(10):123
Jones, Clay (University of New Mexico/USA) *6*(9):223
Jones, Triana N. (University of Maryland/USA) *6*(3):205
Jonker, Hendrikus (Vrije Universiteit/THE NETHERLANDS) *6*(4):91
Ju, Lu-Kwang (The University of Akron/USA) *6*(6):319

Kaludjerski, Milica (San Diego State University/USA) *6*(10):301
Kamashwaran, S. Ramanathen (University of Idaho/USA) *6*(3):91
Kambhampati, Murty S. (Southern University at New Orleans/USA) *6*(5):145, 151
Kamimura, Daisuke (Gunma University/JAPAN) *6*(8):113
Kang, James J. (URS Corporation/USA) *6*(1):121; *6*(10):223
Kappelmeyer, Uwe (UFZ Centre for Environmental Research/GERMANY) *6*(5):337
Karamanev, Dimitre G. (University of Western Ontario/CANADA) *6*(10):171
Karlson, Ulrich (National Environmental Research Institute) *6*(3):141
Kastner, James R. (University of Georgia/USA) *6*(9):289
Kästner, Matthias (UFZ Centre for Environmental Research/GERMANY) *6*(4):99; *6*(5):337
Katz, Lynn E. (University of Texas/USA) *6*(8):139
Kavanaugh, Rathi G. (University of Cincinnati/USA) *6*(2):247

Author Index

171

Kawahara, Fred (U.S. EPA/USA) 6(9):9
Kawakami, Tsuyoshi (University of Tsukuba/JAPAN) 6(9):187
Keefer, Donald A. (Illinois State Geological Survey/USA) 6(9):179
Keith, Nathaniel (Texas A&M University/USA) 6(5):25
Kelly, Laureen S. (Montana Department of Environmental Quality/USA) 6(6):287
Kempisty, David M. (U.S. Air Force/USA) 6(10):145, 155
Kerfoot, William B. (K-V Associates, Inc./USA) 6(10):33
Keuning, S. (Bioclear Environmental Technology/THE NETHERLANDS) 6(8):11
Khan, Tariq A. (Groundwater Services, Inc./USA) 6(7):19
Khodadoust, Amid P. (University of Cincinnati/USA) 6(5):243, 253, 261
Kieft, Thomas L. (New Mexico Institute of Mining and Technology/USA) 6(8):131
Kiessig, Gunter (WISMUT GmbH/GERMANY) 6(5):173; 6(9):155
Kilbride, Rebecca (CEFAS Laboratory/UK) 6(10):337
Kim, Jae Young (Seoul National University/REPUBLIC OF KOREA) 6(9):195
Kim, Jay (University of Cincinnati/USA) 6(6):133
Kim, Kijung (The Pennsylvania State University/USA) 6(9):303
Kim, Tae Young (Ewha University/REPUBLIC OF KOREA) 6(6):51
Kinsall, Barry L. (Oak Ridge National Laboratory/USA) 6(4):73
Kirschenmann, Kyle (IT Corp/USA) 6(4):109
Klaas, Norbert (University of Stuttgart/GERMANY) 6(2):137
Klecka, Gary M. (The Dow Chemical Company/USA) 6(2):89
Klein, Katrina (GeoTrans, Inc./USA) 6(2):163

Klens, Julia L. (IT Corporation/USA) 6(2):253; 6(9):215
Knotek-Smith, Heather M. (University of Idaho/USA) 6(9):147
Koch, Stacey A. (RMT, Inc./USA) 6(7):181
Koenen, Brent A. (U.S. Army Engineer Research & Development Center/USA) 6(5):9
Koenigsberg, Stephen S. (Regenesis Bioremediation Products/USA) 6(7):197, 257; 6(8):209; 6(10):9, 87
Kohata, Kunio (National Institute for Environmental Studies/JAPAN) 6(5):321
Kohler, Keisha (ThermoRetec Corporation/USA) 6(7):1
Kolhatkar, Ravindra V. (BP Corporation/USA) 6(1):35, 43
Komlos, John (Montana State University/USA) 6(7):117
Komnitsas, Kostas (National Technical University of Athens/GREECE) 6(9):97
Kono, Masakazu (Oppenheimer Biotechnology/JAPAN) 6(10):345
Koons, Brad W. (Leggette, Brashears & Graham, Inc./USA) 6(1):175
Koschal, Gerard (PNG Environmental/USA) 6(1):203
Koschorreck, Matthias (UFZ Centre for Environmental Research/GERMANY) 6(9):43
Koshikawa, Hiroshi (National Institute for Environmental Studies/JAPAN) 6(5):321
Kramers, Jan D. (University of Bern/SWITZERLAND) 6(4):91
Krooneman, Jannneke (Bioclear Environmental Biotechnology/THE NETHERLANDS) 6(7):141
Kruk, Taras B. (URS Corporation/USA) 6(10):223
Kuhwald, Jerry (NOVA Chemicals Corporation/CANADA) 6(5):53
Kuschk, Peter (UFZ Centre for Environmental Research Leipzig/GERMANY) 6(5):337

Laboudigue, Agnes (Centre National de Recherche sur les Sites et Sols Pollués/FRANCE) 6(2):129
LaFlamme, Brian (Engineering Management Support, Inc./USA) 6(10):231
Lafontaine, Chantal (École Polytechnique de Montréal/CANADA) 6(10):171
Laha, Shonali (Florida International University/USA) 6(10):187
Laing, M.D. (University of Natal/REP OF SOUTH AFRICA) 6(9):79
Lamar, Richard (EarthFax Development Corp/USA) 6(6):263
Lamarche, Philippe (Royal Military College of Canada/CANADA) 6(8):95
Lamb, Steven R. (GZA GeoEnvironmental, Inc./USA) 6(7):165
Landis, Richard C. (E.I. du Pont de Nemours & Company/USA) 6(8):185
Lang, Beth (United Technologies Corp./USA) 6(10):41
Langenhoff, Alette (TNO Institute of Environmental Science/THE NETHERLANDS) 6(7):141
LaPat-Polasko, Laurie T. (Parsons Engineering Science, Inc./USA) 6(2):65, 189
Lapus, Kevin (Regenesis/USA) 6(7):257; 6(10):9
LaRiviere, Daniel (Texas A&M University/USA) 6(5):17, 25
Larsen, Lars C. (Hedeselskabet/DENMARK) 6(2):11
Larson, John R. (TranSystems Corporation/USA) 6(7):229
Larson, Richard A. (University of Illinois at Urbana-Champaign/USA) 6(5):181
Lauzon, Francois (Dept of National Defence/CANADA) 6(8):95
Leavitt, Maureen E. (Newfields Inc./USA) 6(1):51; 6(5):189
Lebron, Carmen A. (U.S. Navy/USA) 6(7):95
Lee, B. J. (Science Applications International Corporation) 6(8):27

Lee, Brady D. (Idaho National Engineering & Environmental Laboratory/USA) 6(7):77
Lee, Chi Mei (National Chung Hsing University/TAIWAN) 6(6):353
Lee, Eun-Ju (Louisiana State University/USA) 6(5):313
Lee, Kenneth (Fisheries & Oceans Canada/CANADA) 6(10):337
Lee, Michael D. (Terra Systems, Inc./USA) 6(7):213, 249
Lee, Ming-Kuo (Auburn University/USA) 6(9):105
Lee, Patrick (Queen's University/CANADA) 6(2):121
Lee, Seung-Bong (University of Washington/USA) 6(10):211
Lee, Si-Jin (Kyonggi University/REPUBLIC OF KOREA) 6(1):161
Lee, Sung-Jae (ChoongAng University/REPUBLIC OF KOREA) 6(6):51
Leeson, Andrea (Battelle/USA) 6(10):1, 145, 155, 193
Lehman, Stewart E. (California Polytechnic State University/USA) 6(2):1
Lei, Li (University of Cincinnati/USA) 6(5):243, 261
Leigh, Daniel P. (IT Corporation/USA) 6(3):35
Leigh, Mary Beth (University of Oklahoma/USA) 6(5):61
Lendvay, John (University of San Francisco/USA) 6(8):19
Lenzo, Frank C. (ARCADIS Geraghty & Miller/USA) 6(7):53
Leon, Nidya (PDVSA Intevep/VENEZUELA) 6(6):257
Leong, Sylvia (Crescent Heights High School/CANADA) 6(5):53
Leontievsky, Alexey A. (Russian Academy of Sciences/RUSSIA) 6(3):75
Lerner, David N. (University of Sheffield/UK) 6(1):59; 6(2):207
Lesage, Suzanne (National Water Research Institute/CANADA) 6(7):321, 333, 341

Leslie, Jolyn C. (Camp Dresser & McKee, Inc./USA) 6(2):113
Lewis, Ronald F. (U.S. EPA/USA) 6(5):253, 261
Li, Dong X. (USA) 6(7):205
Li, Guanghe (Tsinghua University/CHINA) 6(7):61
Li, Tong (Tetra Tech EM Inc./USA) 6(10):293
Librando, Vito (Universita di Catania/ITALY) 6(3):149
Lieberman, M. Tony (Solutions Industrial & Environmental Services/USA) 6(7):249
Lin, Cindy (Conestoga-Rovers & Associates/USA) 6(1):99; 6(10):131
Lipson, David S. (Blasland, Bouck & Lee, Inc./USA) 6(10):319
Liu, Jian (University of Nevada Las Vegas/USA) 6(9):265
Liu, Xiumei (Shandong Agricultural University/ CHINA) 6(9):113
Livingstone, Stephen (Franz Environmental Inc./CANADA) 6(6):211
Lizzari, Daniela (Universita degli Studi di Verona/ITALY) 6(3):267
Llewellyn, Tim (URS/USA) 6(7):321, 333, 341
Lobo, C. (El Encin IMIA/SPAIN) 6(4):29
Loeffler, Frank E. (Georgia Institute of Technology/USA) 6(8):19
Logan, Bruce E. (The Pennsylvania State University/USA) 6(9):303
Long, Gilbert M. (Camp Dresser & McKee Inc./USA) 6(6):287
Longoni, Giovanni (Montgomery Watson/ITALY) 6(10):41
Lorbeer, Helmut (Technical University of Dresden/GERMANY) 6(8):105
Lors, Christine (Centre National de Recherche sur les Sites et Sols Pollués /FRANCE) 6(2):129
Lorton, Diane M. (King's College London/UK) 6(2):223; 6(3):173
Losi, Mark E. (Foster Wheeler Environ. Corp./USA) 6(9):249
Loucks, Mark (U.S. Air Force/USA) 6(2):261

Lu, Chih-Jen (National Chung Hsing University/TAIWAN) 6(6):353; 6(10):217
Lu, Xiaoxia (Tsinghua University/CHINA) 6(7):61
Lubenow, Brian (University of Delaware/USA) 6(6):149
Lucas, Mary (Parsons Engineering Science, Inc./USA) 6(10):283
Lundgren, Tommy S. (Sydkraft SAKAB AB/SWEDEN) 6(6):127
Lundstedt, Staffan (Umeå University/SWEDEN) 6(3):181
Luo, Xiaohong (NRC Research Associate/USA) 6(8):167
Luthy, Richard G. (Stanford University/USA) 6(3):189
Lutze, Werner (University of New Mexico/USA) 6(9):155
Luu, Y.-S. (Queen's University/CANADA) 6(2):121
Lynch, Regina M. (Battelle/USA) 6(10):155

Macek, Thomáš (Institute of Chemical Technology/Czech Republic) 6(5):61
MacEwen, Scott J. (CH2M Hill/USA) 6(3):107
Machado, Kátia M. G. (Fund. Centro Tecnológico de Minas Gerais/BRAZIL) 6(3):99
Maciel, Helena Alves (Aberdeen University/UK) 6(1):1
Mack, E. Erin (E.I. du Pont de Nemours & Co./USA) 6(2):81; 6(8):43
Macková, Martina (Institute of Chemical Technology/Czech Republic) 6(5):61
Macnaughton, Sarah J. (AEA Technology/UK) 6(5):305; 6(10):337
Macomber, Jeff R. (University of Cincinnati/USA) 6(6):133
Macrae, Jean (University of Maine/USA) 6(8):139
Madden, Patrick C. (Engineering Consultant/USA) 6(8):87
Madsen, Clint (Terracon/USA) 6(8):157; 6(10):15
Magar, Victor S. (Battelle/USA) 6(1):137; 6(5):231, 237; 6(10):145, 155

Mage, Roland (Battelle Europe/SWITZERLAND) 6(6):241; 6(10):109
Magistrelli, P. (Istituto di Cibernetica e Biofisica/ITALY) 6(5):157
Maierle, Michael S. (ARCADIS Geraghty & Miller, Inc./USA) 6(7):149
Major, C. Lee (Jr.) (University of Michigan/USA) 6(8):19
Major, David W. (GeoSyntec Consultants/CANADA) 6(8):27
Maki, Hideaki (National Institute for Environmental Studies/JAPAN) 6(5):321
Makkar, Randhir S. (University of Illinois-Chicago/USA) 6(5):297
Malcolm, Dave (BAE Systems Properties Ltd./UK) 6(6):119
Manabe, Takehiko (Hyogo Prefectural Fisheries Research Institute/JAPAN) 6(10):345
Maner, P.M. (Equilon Enterprises, LLC/USA) 6(1):11
Maner, Paul (Shell Development Company/USA) 6(8):61
Manrique-Ramírez, Emilio Javier (SYMCA, S.A. de C.V./MEXICO) 6(6):369
Marchal, Rémy (Institut Français du Pétrole/FRANCE) 6(1):153
Maresco, Vincent (Groundwater & Environmental Srvcs/USA) 6(10):101
Marnette, Emile C. (TAUW BV/THE NETHERLANDS) 6(7):297
Marshall, Timothy R. (URS Corporation/USA) 6(2):49
Martella, L. (Istituto di Cibernetica e Biofisica/ITALY) 6(5):157
Martin, C. (Universidad Politecnica/SPAIN) 6(4):29
Martin, Jennifer P. (Idaho National Engineering & Environmental Laboratory/USA) 6(7):265
Martin, John F. (U.S. EPA/USA) 6(2):247
Martin, Margarita (Universidad Complutense de Madrid/SPAIN) 6(4):29; 6(6):377

Martinez-Inigo, M.J. (El Encin IMIA/SPAIN) 6(4):29
Martino, Lou (Argonne National Laboratory/USA) 6(5):207
Mascarenas, Tom (Environmental Chemistry/USA) 6(8):157
Mason, Jeremy (King's College London/UK) 6(2):223; 6(3):173; 6(10):123
Massella, Oscar (Universita degli Studi di Verona/ITALY) 6(3):267
Matheus, Dacio R. (Instituto de Botânica/BRAZIL) 6(3):99
Matos, Tania (University of Puerto Rico at Rio Piedras/USA) 6(9):179
Matsubara, Takashi (Obayashi Corporation/JAPAN) 6(6):249
Mattiasson, Bo (Lund University/SWEDEN) 6(3):181; 6(6):65; 6(9):123
McCall, Sarah (Battelle/USA) 6(10):155, 245
McCarthy, Kevin (Battelle Duxbury Operations/USA) 6(5):9
McCartney, Daryl M. (University of Manitoba/CANADA) 6(6):165
McCormick, Michael L. (The University of Michigan/USA) 6(8):193
McDonald, Thomas J. (Texas A&M University) 6(5):17
McElligott, Mike (U.S. Air Force/USA) 6(1):51
McGill, William B. (University of Northern British Columbia/CANADA) 6(4):7
McIntosh, Heather (U.S. Army/USA) 6(7):321, 333
McLinn, Eugene L. (RMT, Inc./USA) 6(5):121
McLoughlin, Patrick W. (Microseeps Inc./USA) 6(1):35
McMaster, Michaye (GeoSyntec Consultants/CANADA) 6(8):27, 43, 6(9):297
McMillen, Sara J. (Chevron Research & Technology Company/USA) 6(6):193
Meckenstock, Rainer U. (University of Tübingen/GERMANY) 6(4):99
Mehnert, Edward (Illinois State Geological Survey/USA) 6(9):179

Meigio, Jodette L. (Idaho National Engineering & Environmental Laboratory/USA) 6(7):77
Meijer, Harro A.J. (University of Groningen/THE NETHERLANDS) 6(4):91
Meijerink, E. (Province of Drenthe/THE NETHERLANDS) 6(8):11
Merino-Castro, Glicina (Inst Technol y de Estudios Superiores/MEXICO) 6(6):377
Messier, J.P. (U.S. Coast Guard/USA) 6(1):107
Meyer, Michael (Environmental Resources Management/BELGIUM) 6(7):87
Meylan, S. (Queen's University/CANADA) 6(2):121
Miles, Victor (Duracell Inc./USA) 6(7):87
Millar, Kelly (National Water Research Institute/CANADA) 6(7):321, 333, 341
Miller, Michael E. (Camp Dresser & McKee, Inc./USA) 6(7):273
Miller, Thomas Ferrell (Lockheed Martin/USA) 6(3):259
Mills, Heath J. (Georgia Institute of Technology/USA) 6(9):165
Millward, Rod N. (Louisiana State University/USA) 6(5):305
Mishra, Pramod Chandra (Sambalpur University/INDIA) 6(9):173
Mitchell, David (AEA Technology Environment/UK) 6(10):337
Mitraka, Maria (Serres/GREECE) 6(6):89
Mocciaro, PierFilippo (Ambiente S.p.A./ITALY) 6(6):227
Moeri, Ernesto N. (CSD-GEOKLOCK/BRAZIL) 6(1):27
Moir, Michael (Chevron Research & Technology Co./USA) 6(1):83
Molinari, Mauro (AgipPetroli S.p.A/ITALY) 6(6):173
Mollea, C. (Politecnico di Torino/ITALY) 6(3):211
Mollhagen, Tony (Texas Tech University/USA) 6(3):9
Monot, Frédéric (Institut Français du Pétrole/FRANCE) 6(1):153

Moon, Hee Sun (Seoul National University/REPUBLIC OF KOREA) 6(9):195
Moosa, Shehnaaz (University of Cape Town/REP OF SOUTH AFRICA) 6(6):339
Morasch, Barbara (University Konstanz/GERMANY) 6(4):99
Moreno, Joanna (URS Corporation/USA) 6(2):239
Morgan, Scott (URS - Dames & Moore/USA) 6(7):321
Morrill, Pamela J. (Camp, Dresser, & McKee, Inc./USA) 6(2):113
Morris, Damon (ThermoRetec Corporation/USA) 6(7):1
Mortimer, Marylove (Mississippi State University/USA) 6(5):329
Mortimer, Wendy (Bell Canada/CANADA) 6(2):27; 6(6):185, 203, 211,
Mossing, Christian (Hedeselskabet/DENMARK) 6(2):11
Mossmann, Jean-Remi (Centre National de Recherche sur les Sites et Sols Pollués/FRANCE) 6(2):129
Moteleb, Moustafa A. (University of Cincinnati/USA) 6(6):133
Mowder, Carol S. (URS/USA) 6(7):321, 333, 341
Moyer, Ellen E. (ENSR International./USA) 6(1):75
Mravik, Susan C. (U.S. EPA/USA) 6(1):167
Mueller, James G. (URS Corporation/USA) 6(2):239
Müller, Axel (Water Technology Center/GERMANY) 6(8):105
Müller, Beate (Umweltschutz Nord GmbH/GERMANY) 6(4):131
Müller, Klaus (Battelle Europe/SWITZERLAND) 6(5):41; 6(6):241
Muniz, Herminio (Hart Crowser Inc./USA) 6(10):9
Murphy, Sean M. (Komex International Ltd./CANADA) 6(4):7
Murray, Cliff (United States Army Corps of Engineers/USA) 6(9):281
Murray, Gordon Bruce (Stella-Jones Inc./CANADA) 6(3):197

Murray, Willard A. (Harding ESE/USA) 6(7):197
Mutch, Robert D. (Brown and Caldwell/USA) 6(2):145
Mutti, Francois (Water & Soil Remediation S.r.l./ITALY) 6(6):179
Myasoedova, Nina M. (Russian Academy of Sciences/RUSSIA) 6(3):75

Nadolishny, Alex (Nedatek, Inc./USA) 6(10):139
Nagle, David P. (University of Oklahoma/USA) 6(5):61
Nam, Kyoungphile (Seoul National University/REPUBLIC OF KOREA) 6(9):195
Narayanaswamy, Karthik (Parsons Engineering Science/USA) 6(2):65
Nelson, Mark D. (Delta Environmental Consultants, Inc./USA) 6(1):175
Nelson, Yarrow (California Polytechnic State University/USA) 6(10):311
Nemati, M. (University of Cape Town/REP OF SOUTH AFRICA) 6(6):339
Nestler, Catherine C. (Applied Research Associates, Inc./USA) 6(4):59, 6(6):43
Nevárez-Moorillón, G.V. (UACH/MEXICO) 6(6):361
Neville, Scott L. (Aerojet General Corp./USA) 6(9):297
Newell, Charles J. (Groundwater Services, Inc./USA) 6(7):19
Nieman, Karl (Utah State University/USA) 6(4):67
Niemeyer, Thomas (Hochtief Umwelt Gmbh/GERMANY) 6(6):227
Nies, Loring (Purdue University/USA) 6(3):125
Nipshagen, Adri A.M. (IWACO/THE NETHERLANDS) 6(7):141
Nishino, Shirley (U.S. Air Force/USA) 6(3):59
Nivens, David E. (University of Tennessee/USA) 6(4):45
Noffsinger, David (Westinghouse Savannah River Company/USA) 6(10):163

Noguchi, Takuya (Nippon Institute of Technology/JAPAN) 6(5):321
Nojiri, Hideaki (The University of Tokyo/JAPAN) 6(4):51; 6(6):111
Noland, Scott (NESCO Inc./USA) 6(10):73
Nolen, C. Hunter (Camp Dresser & McKee/USA) 6(6):287
Norris, Robert D. (Eckenfelder/Brown and Caldwell/USA) 6(2):145; 6(7):35
North, Robert W. (Environ Corporation./USA) 6(7):189
Novak, John T. (Virginia Polytechnic Institute & State University/USA) 6(2):105; 6(5):1
Novick, Norman (Exxon/Mobil Oil Corp/USA) 6(1):35
Nuttall, H. Eric (The University of New Mexico/USA) 6(9): 155, 223
Nuyens, Dirk (Environmental Resources Management/BELGIUM) 6(7):87; 6(9):87
Nzengung, Valentine A. (University of Georgia/USA) 6(9):289

Ochs, L. Donald (Regenesis/USA) 6(10):139
O'Connell, Joseph E. (Environmental Resolutions, Inc./USA) 6(1):91
Odle, Bill (Newfields, Inc./USA) 6(5):189
O'Donnell, Ingrid (BAE Systems Properties, Ltd./UK) 6(6):119
Ogden, Richard (BAE Systems Properties Ltd./UK) 6(6):119
Oh, Byung-Taek (The University of Iowa/USA) 6(8):147, 175
Oh, Seok-Young (University of Delaware/USA) 6(6):149
Omori, Toshio (The University of Tokyo/JAPAN) 6(4):51; 6(6):111
O'Neal, Brenda (ARA/USA) 6(3):43
Oppenheimer, Carl H. (Oppenheimer Biotechnology/USA) 6(10):345
O'Regan, Gerald (Chevron Products Company/USA) 6(1):203
O'Reilly, Kirk T. (Chevron Research & Technology Co/USA) 6(1):83, 145, 203
Oshio, Takahiro (University of Tsukuba/JAPAN) 6(9):187

Ozdemiroglu, Ece (EFTEC Ltd./UK) 6(4):17

Padovani, Marco (Centro Ricerche Ambientali/ITALY) 6(4):131

Paganetto, A. (Istituto di Cibernetica e Biofisica/ITALY) 6(5):157

Pahr, Michelle R. (ARCADIS Geraghty & Miller/USA) 6(1):107

Pal, Nirupam (California Polytechnic State University/USA) 6(2):1

Palmer, Tracy (Applied Power Concepts, Inc./USA) 6(7):103

Palumbo, Anthony V. (Oak Ridge National Laboratory/USA) 6(4):73; 6(9):165

Panciera, Matthew A. (University of Connecticut/USA) 6(7):69

Pancras, Tessa (Wageningen University/THE NETHERLANDS) 6(5):289

Pardue, John H. (Louisiana State University/USA) 6(5): 207, 313; 6(9):273

Park, Kyoohong (ChoongAng University/REPUBLIC OF KOREA) 6(6):51

Parkin, Gene F. (University of Iowa/USA) 6(3):1

Paspaliaris, Ioannis (National Technical University of Athens/GREECE) 6(9):97

Paton, Graeme I. (Aberdeen University/UK) 6(1):1

Patrick, John (University of Reading/UK) 6(10):337

Payne, Frederick C. (ARCADIS Geraghty & Miller/USA) 6(7):53

Payne, Jo Ann (DuPont Co./USA) 6(8):43

Peabody, Jack G. (Regenesis/USA) 6(10):95

Peacock, Aaron D. (University of Tennessee/USA) 6(4):73; 6(5):305

Peargin, Tom R. (Chevron Research & Technology Co/USA) 6(1):67

Peeples, James A. (Metcalf & Eddy, Inc./USA) 6(7):173

Pehlivan, Mehmet (Tait Environmental Management, Inc./USA) 6(10):267, 275

Pelletier, Emilien (ISMER/CANADA) 6(2):57

Pennie, Kimberley A. (Stella-Jones, Inc./CANADA) 6(3):197

Peramaki, Matthew P. (Leggette, Brashears, & Graham, Inc./USA) 6(10):259

Perey, Jennie R. (University of Delaware/USA) 6(6):149

Perez-Vargas, Josefina (CINVESTAV-IPN/MEXICO) 6(6):219

Perina, Tomas (IT Corporation/USA) 6(1):51; 6(2):73

Perlis, Shira R. (Rowan University/USA) 6(5):215

Perlmutter, Michael W. (EnSafe, Inc./USA) 6(9):315

Perrier, Michel (École Polytechnique de Montréal/CANADA) 6(4):139

Perry, L.B. (U.S. Army Engineer Research & Development Center/USA) 6(5):9

Persico, John L. (Blasland, Bouck & Lee, Inc./USA) 6(10):319

Peschong, Bradley J. (Leggette, Brashears & Graham, Inc./USA) 6(1):175

Peters, Dave (URS/USA) 6(7):333

Peterson, Lance N. (North Wind Environmental, Inc./USA) 6(7):265

Petrovskis, Erik A. (Geotrans Inc./USA) 6(8):19

Peven-McCarthy, Carole (Battelle Ocean Sciences/USA) 6(5):231

Pfiffner, Susan M. (University of Tennessee/USA) 6(4):73

Phelps, Tommy J. (Oak Ridge National Laboratory/USA) 6(4):73

Pickett, Tim M. (Applied Biosciences Corporation/USA) 6(9):331

Pickle, D.W. (Equilon Enterprises LLC/USA) 6(8):61

Pierre, Stephane (École Polytechnique de Montréal/CANADA) 6(10):171

Pijls, Charles G.J.M. (TAUW BV/THE NETHERLANDS) 6(10):253

Pirkle, Robert J. (Microseeps, Inc./USA) 6(1):35

Pisarik, Michael F. (New Fields/USA) 6(1):121

Piveteau, Pascal (Institut Français du Pétrole/FRANCE) 6(1):153
Place, Matthew (Battelle/USA) 6(10):245
Plata, Nadia (Battelle Europe/SWITZERLAND) 6(5):41
Poggi-Varaldo, Hector M. (CINVESTAV-IPN/MEXICO) 6(3):243; 6(6):219
Pohlmann, Dirk C. (IT Corporation/USA) 6(2):253
Pokethitiyook, Prayad (Mahidol University/THAILAND) 6(10):329
Polk, Jonna (U.S. Army Corps of Engineers/USA) 6(9):281
Pope, Daniel F. (Dynamac Corp/USA) 6(1):129
Porta, Augusto (Battelle Europe/SWITZERLAND) 6(5):41; 6(6):241; 6(10):109
Portier, Ralph J. (Louisiana State University/USA) 6(5):305
Powers, Leigh (Georgia Institute of Technology/USA) 6(9):165
Prandi, Alberto (Water & Soil Remediation S.r.l/ITALY) 6(6):179
Prasad, M.N.V. (University of Hyderabad/INDIA) 6(5):165
Price, Steven (Camp Dresser & McKee, Inc./USA) 6(9):303
Priester, Lamar E. (Priester & Associates/USA) 6(10):65
Pritchard, P. H. (Hap) (U.S. Navy/USA) 6(7):125
Profit, Michael D. (CDM Federal Programs Corporation/USA) 6(6):81
Prosnansky, Michal (Gunma University/JAPAN) 6(9):201
Pruden, Amy (University of Cincinnati/USA) 6(1):19
Ptacek, Carol J. (University of Waterloo/CANADA) 6(9):71

Radosevich, Mark (University of Delaware/USA) 6(6):9
Radtke, Corey (INEEL/USA) 6(3):9
Raetz, Richard M. (Global Remediation Technologies, Inc./USA) 6(6):311
Rainwater, Ken (Texas Tech University/USA) 6(3):9

Ramani, Mukundan (University of Cincinnati/USA) 6(5):269
Raming, Julie B. (Georgia-Pacific Corp./USA) 6(1):183
Ramírez, N. E. (ECOPETROL-ICP/COLOMBIA) 6(6):319
Ramsay, Bruce A. (Polyferm Canada Inc./CANADA) 6(2):121; 6(10):171
Ramsay, Juliana A. (Queen's University/CANADA) 6(2):121; 6(10):171
Rao, Prasanna (University of Cincinnati/USA) 6(9):1
Ratzke, Hans-Peter (Umweltschutz Nord GMBH/GERMANY) 6(4):131
Reardon, Kenneth F. (Colorado State University/USA) 6(8):53
Rectanus, Heather V. (Virginia Polytechnic Institute & State University/USA) 6(2):105
Reed, Thomas A. (URS Corporation/USA) 6(8):157; 6(10):15, 95
Rees, Hubert (CEFAS Laboratory/UK) 6(10):337
Rehm, Bernd W. (RMT, Inc./USA) 6(2):97; 6(10):201
Reinecke, Stefan (Franz Environmental Inc./CANADA) 6(6):211
Reinhard, Martin (Stanford University/USA) 6(7):95
Reisinger, H. James (Integrated Science & Technology Inc/USA) 6(1):183
Rek, Dorota (IT Corporation/USA) 6(2):73
Reynolds, Charles M. (U.S. Army Engineer Research & Development Center/USA) 6(5):9
Reynolds, Daniel E. (Air Force Institute of Technology/USA) 6(2):173
Rice, John M. (RMT, Inc./USA) 6(7):181
Richard, Don E. (Barr Engineering Company/USA) 6(3):219; 6(5):105
Richardson, Ian (Conestoga-Rovers & Associates/USA) 6(10):131
Richnow, Hans H. (UFZ-Centre for Environmental Research/GERMANY) 6(4):99

Rijnaarts, Huub H.M. (TNO Institute of Environmental Science/THE NETHERLANDS) 6(2):231
Ringelberg, David B. (U.S. Army Corps of Engineers/USA) 6(5):9; 6(6):43; 6(10):115
Ríos-Leal, E. (CINVESTAV-IPN/MEXICO) 6(3):243
Ripp, Steven (University of Tennessee/USA) 6(4):45
Ritter, Michael (URS Corporation/USA) 6(2):239
Ritter, William F. (University of Delaware/USA) 6(6):9
Riva, Vanessa (Parsons Engineering Science, Inc./USA) 6(2):39
Rivas-Lucero, B.A. (Centro de Investigacion en Materiales Avanzados/MEXICO) 6(6):361
Rivetta, A. (Universita degli Studi di Milano/ITALY) 6(5):157
Robb, Joseph (ENSR International/USA) 6(1):75
Robertiello, Andrea (EniTecnologie S.p.A./ITALY) 6(6):173
Robertson, K. (Queen's University/CANADA) 6(2):121
Robinson, David (ERM, Inc./USA) 6(7):45
Robinson, Sandra L. (Virginia Polytechnic Institute & State University/USA) 6(5):1
Rockne, Karl J. (University of Illinois-Chicago/USA) 6(5):297
Rodríguez-Vázquez, Refugio (CINVESTAV-IPN/MEXICO) 6(3):243; 6(6):219, 369
Römkens, Paul (Alterra/THE NETHERLANDS) 6(5):137
Rongo, Rocco (University della Calabria/ITALY) 6(4):131
Roorda, Marcus L. (Rowan University/USA) 6(5):215
Rosser, Susan J. (University of Cambridge/UK) 6(5):69
Rowland, Martin A. (Lockheed-Martin Michoud Space Systems/USA) 6(7):1
Royer, Richard (The Pennsylvania State University/USA) 6(8):201

Ruggeri, Bernardo (Politecnico di Torino/ITALY) 6(3):211
Ruiz, Graciela M. (University of Iowa/USA) 6(1):195
Rupassara, S. Indumathie (University of Illinois at Urbana-Champaign/USA) 6(5):181

Sacchi, G.A. (Universita degli Studi di Milano/ITALY) 6(5):157
Sahagun, Tracy (U.S. Marine Corps./USA) 6(10):1
Sakakibara, Yutaka (Waseda University/JAPAN) 6(8):113; 6(9):201
Sakamoto, T. (Queen's University/CANADA) 6(10):171
Salam, Munazza (Crescent Heights High School/CANADA) 6(5):53
Salanitro, Joseph P. (Equilon Enterprises, LLC/USA) 6(1):11; 6(8):61
Salvador, Maria Cristina (CSD-GEOKLOCK/BRAZIL) 6(1):27
Samson, Réjean (École Polytechnique de Montréal/CANADA) 6(3):115; 6(4):139; 6(9):27
San Felipe, Zenaida (Monash University/AUSTRALIA) 6(4):1
Sánchez, F.N. (ECOPETROL-ICP/COLOMBIA) 6(6):319
Sánchez, Gisela (PDVSA Intevep/VENEZUELA) 6(6):257
Sánchez, Luis (PDVSA Intevep/VENEZUELA) 6(6):257
Sanchez, M. (Universidad Complutense de Madrid/SPAIN) 6(4):29; 6(6):377
Sandefur, Craig A. (Regenesis/USA) 6(7):257; 6(10):87
Sanford, Robert A. (University of Illinois at Urbana-Champaign/USA) 6(9):179
Santangelo-Dreiling, Theresa (Colorado Dept. of Transportation/USA) 6(10):231
Saran, Jennifer (Kennecott Utah Copper Corp./USA) 6(9):323
Sarpietro, M.G. (Universita di Catania/ITALY) 6(3):149

Sartoros, Catherine (Université du Québec à Montréal/CANADA) 6(3):165
Saucedo-Terán, R.A. (Centro de Investigacion en Materiales Avanzados/MEXICO) 6(6):361
Saunders, James A. (Auburn University/USA) 6(9):105
Sayler, Gary S. (University of Tennessee/USA) 6(4):45
Scalzi, Michael M. (Innovative Environmental Technologies, Inc./USA) 6(10):23
Scarborough, Shirley (IT Corporation/USA) 6(2):253
Schaffner, I. Richard (GZA GeoEnvironmental, Inc./USA) 6(7):165
Scharp, Richard A. (U.S. EPA/USA) 6(9):9
Schell, Heico (Water Technology Center/GERMANY) 6(8):105
Scherer, Michelle M. (The University of Iowa/USA) 6(3):1
Schipper, Mark (Groundwater Services) 6(8):73
Schmelling, Stephen (U.S. EPA/USA) 6(1):129
Schnoor, Jerald L. (University of Iowa/USA) 6(8):147
Schoefs, Olivier (École Polytechnique de Montréal/CANADA) 6(4):139
Schratzberger, Michaela (CEFAS Laboratory/UK) 6(10):337
Schulze, Susanne (Water Technology Center/GERMANY) 6(2):137
Schuur, Jessica H. (Lund University/SWEDEN) 6(6):65
Scrocchi, Susan (Conestoga-Rovers & Associates/USA) 6(1):99; 6(10):131
Sczechowski, Jeff (California Polytechnic State University/USA) 6(10):311
Seagren, Eric A. (University of Maryland/USA) 6(10):57
Sedran, Marie A. (University of Cincinnati/USA) 6(1):19
Seifert, Dorte (Technical University of Denmark/DENMARK) 6(2):11
Semer, Robin (Harza Engineering Company, Inc./USA) 6(7):157

Semprini, Lewis (Oregon State University/USA) 6(10):145, 155, 179
Seracuse, Joe (Harding ESE/USA) 6(7):205
Serra, Roberto (Centro Ricerche Ambientali/ITALY) 6(4):131
Sewell, Guy W. (U.S. EPA/USA) 6(1):167; 6(7):125; 6(8):167
Sharma, Pawan (Camp Dresser & McKee Inc./USA) 6(7):305
Sharp, Robert R. (Manhattan College/USA) 6(7):117
Shay, Devin T. (Groundwater & Environmental Services, Inc./USA) 6(10):101
Shelley, Michael L. (Air Force Institute of Technology/USA) 6(5):95
Shen, Hai (Dynamac Corporation/USA) 6(1): 129, 167
Sherman, Neil (Louisiana-Pacific Corporation/USA) 6(3):83
Sherwood Lollar, Barbara (University of Toronto/CANADA) 6(4):91, 109
Shi, Jing (EFX Systems, Inc./USA) 6(8):19
Shields, Adrian R.G. (Komex Europe/UK) 6(10):123
Shiffer, Shawn (University of Illinois/USA) 6(9):179
Shin, Won Sik (Lousiana State University/USA) 6(5):313
Shiohara, Kei (Mississippi State University/USA) 6(6):279
Shirazi, Fatemeh R. (Stratum Engineering Inc./USA) 6(8):121
Shoemaker, Christine (Cornell University/USA) 6(4):125
Sibbett, Bruce (IT Corporation/USA) 6(2):73
Silver, Cannon F. (Parsons Engineering Science, Inc./USA) 6(10):283
Silverman, Thomas S. (RMT, Inc./USA) 6(10):201
Simon, Michelle A. (U.S. EPA/USA) 6(10):293
Sims, Gerald K. (USDA-ARS/USA) 6(5):181
Sims, Ronald C. (Utah State University/USA) 6(4):67; 6(6):1
Sincock, M. Jennifer (ENVIRON International Corp./USA) 6(7):189

Sittler, Steven P. (Advanced Pollution Technologists, Ltd./USA) 6(2):215
Skladany, George J. (ERM, Inc./USA) 6(7):45, 213
Skubal, Karen L. (Case Western Reserve University/USA) 6(8):193
Slenders, Hans (TNO-MEP/THE NETHERLANDS) 6(7):289
Slomczynski, David J. (University of Cincinnati/USA) 6(2):247
Slusser, Thomas J. (Wright State University/USA) 6(5):95
Smallbeck, Donald R. (Harding Lawson/USA) 6(10):231
Smets, Barth F. (University of Connecticut/USA) 6(7):69
Smith, Christy (North Carolina State University/USA) 6(1):145
Smith, Colin C. (University of Sheffield/UK) 6(2):207
Smith, John R. (Alcoa Inc./USA) 6(6):89
Smith, Jonathan (The Environment Agency/UK) 6(4):17
Smith, Steve (King's College London/UK) 6(2):223; 6(3):173; 6(10):123
Smyth, David J.A. (University of Waterloo/CANADA) 6(9):71
Sobecky, Patricia (Georgia Institute of Technology/USA) 6(9):165
Sola, Adrianna (Spelman College/USA) 6(9):53
Sordini, E. (EniTechnologie/ITALY) 6(6):173
Sorensen, James A. (University of North Dakota/USA) 6(6):141
Sorenson, Kent S. (Idaho National Engineering and Environmental Laboratory./USA) 6(7):265
South, Daniel (Harding ESE/USA) 6(7):205
Spain, Jim (U.S. Air Force/USA) 6(3):59; 6(7):125
Spasova, Irena Ilieva (University of Mining & Geology/BULGARIA) 6(9):97
Spataro, William (University della Calabria/ITALY) 6(4):131

Spinnler, Gerard E. (Equilon Enterprises, LLC/USA) 6(1):11; 6(8):61
Springael, Dirk (VITO/BELGIUM) 6(4):35
Srinivasan, P. (GeoTrans, Inc./USA) 6(2):163
Stansbery, Anita (California Polytechnic State University/USA) 6(10):311
Starr, Mark G. (DuPont Co./USA) 6(8):43
Stehmeier, Lester G. (NOVA Research Technology Centre/CANADA) 6(4):117; 6(5):53
Stensel, H. David (University of Washington/USA) 6(10):211
Stordahl, Darrel M. (Camp Dresser & McKee Inc./USA) 6(6):287
Stout, Scott (Battelle/USA) 6(5):237
Strand, Stuart E. (University of Washington/USA) 6(10):211
Stratton, Glenn (Nova Scotia Agricultural College/CANADA) 6(3):197
Strybel, Dan (IT Corporation/USA) 6(9):215
Stuetz, R.M. (Cranfield University/UK) 6(6):329
Suarez, B. (ECOPETROL-ICP/COLOMBIA) 6(6):319
Suidan, Makram T. (University of Cincinnati/USA) 6(1):19; 6(5):243, 253, 261; 6(6):133,
Suthersan, Suthan S. (ARCADIS Geraghty & Miller/USA) 6(7):53
Suzuki, Masahiro (Nippon Institute of Technology/JAPAN) 6(5):321
Sveum, Per (Deconterra AS/NORWAY) 6(6):157
Swallow, Ian (BAE Systems Properties Ltd./UK) 6(6):119
Swann, Benjamin M. (Camp Dresser & McKee Inc./USA) 6(7):305
Swannell, Richard P.J. (AEA Technology Environment/UK) 6(10):337

Tabak, Henry H. (U.S. EPA/USA) 6(5):243, 253, 261, 269; 6(9):1, 17
Takai, Koji (Fuji Packing/JAPAN) 6(10):345

Talley, Jeffrey W. (University of Notre Dame/USA) 6(3):189; 6(4):59; 6(6):43; 6(7):125; 6(10):115
Tao, Shu (Peking University/CHINA) 6(7):61
Taylor, Christine D. (North Carolina State University/USA) 6(1):83
Ter Meer, Jeroen (TNO Institute of Environmental Science/THE NETHERLANDS) 6(2):231; 6(7):289
Tétreault, Michel (Royal Military College of Canada/CANADA) 6(8):95
Tharpe, D.L. (Equilon Enterprises LLC/USA) 6(8):61
Theeuwen, J. (Grontmij BV/THE NETHERLANDS) 6(7):289
Thomas, Hartmut (WASAG DECON GMbH/GERMANY) 6(3):67
Thomas, Mark (EG&G Technical Services, Inc./USA) 6(10):49
Thomas, Paul R. (Thomas Consultants, Inc./USA) 6(5):189
Thomas, Robert C. (University of Georgia/USA) 6(9):105
Thomson, Michelle M. (URS Corporation/USA) 6(2):81
Thornton, Steven F. (University of Sheffield/UK) 6(1):59, 6(2):207
Tian, C. (University of Cincinnati/USA) 6(8):35
Tiedje, James M. (Michigan State University/USA) 6(7):125; 6(8):19
Tiehm, Andreas (Water Technology Center/GERMANY) 6(2):137; 6(8):105
Tietje, David (Foster Wheeler Environmental Corportation/USA) 6(9):249
Timmins, Brian (Oregon State University/USA) 6(10):179
Togna, A. Paul (Envirogen Inc/USA) 6(9):281
Tolbert, David E.(U.S. Army/USA) 6(9):281
Tonnaer, Haimo (TAUW BV/THE NETHERLANDS) 6(7):297; 6(10):253
Toth, Brad (Harding ESE/USA) 6(10):231

Tovanabootr, Adisorn (Oregon State University/USA) 6(10):145
Travis, Bryan (Los Alamos National Laboratory/USA) 6(10):163
Trudnowski, John M. (MSE Technology Applications, Inc./USA) 6(9):35
Truax, Dennis D. (Mississippi State University/USA) 6(9):241
Trute, Mary M. (Camp Dresser & McKee, Inc./USA) 6(2):113
Tsuji, Hirokazu (Obayashi Corporation Ltd./JAPAN) 6(6):111, 249; 6(10):239
Tsutsumi, Hiroaki (Prefectural University of Kumamoto/JAPAN) 6(10):345
Turner, Tim (CDM Federal Programs Corp./USA) 6(6):81
Turner, Xandra (International Biochemicals Group/USA) 6(10):23
Tyner, Larry (IT Corporation/USA) 6(1):51; 6(2):73

Ugolini, Nick (U.S. Navy/USA) 6(10):65
Uhler, Richard (Battelle/USA) 6(5):237
Unz, Richard F. (The Pennsylvania State University/USA) 6(8):201
Utgikar, Vivek P. (U.S. EPA/USA) 6(9):17

Valderrama, Brenda (Universidad Nacional Autónoma de México/MEXICO) 6(6):17
Vallini, Giovanni (Universita degli Studi di Verona/ITALY) 6(3):267
van Bavel, Bert (Umeå University/SWEDEN) 6(3):181
van Breukelen, Boris M. (Vrije University/THE NETHERLANDS) 6(4):91
VanBroekhoven, K. (Catholic University of Leuven/BELGIUM) 6(4):35
Vandecasteele, Jean-Paul (Institut Français du Pétrole/FRANCE) 6(3):227
VanDelft, Frank (NOVA Chemicals/CANADA) 6(5):53
van der Gun, Johan (BodemBeheer bv/THE NETHERLANDS) 6(5):289

van der Werf, A. W. (Bioclear Environmental Technology/THE NETHERLANDS) 6(8):11
van Eekert, Miriam (TNO Environmental Sciences /THE NETHERLANDS) 6(2):231; 6(7):289
Van Hout, Amy H. (IT Corporation/USA) 6(3):35
Van Keulen, E. (DHV Environment and Infrastructure/THE NETHERLANDS) 6(8):11
Vargas, M.C. (ECOPETROL-ICP/COLOMBIA) 6(6):319
Vazquez-Duhalt, Rafael (Universidad Nacional Autónoma de México/MEXICO) 6(6):17
Venosa, Albert (U.S. EPA/USA) 6(1):19
Verhaagen, P. (Grontmij BV/THE NETHERLANDS) 6(7):289
Verheij, T. (DAF/THE NETHERLANDS) 6(7):289
Vidumsky, John E. (E.I. du Pont de Nemours & Company/USA) 6(2):81; 6(8):185
Villani, Marco (Centro Ricerche Ambientali/ITALY) 6(4):131
Vinnai, Louise (Investigative Science Inc./CANADA) 6(2):27
Visscher, Gerolf (Province of Groningen/THE NETHERLANDS) 6(7):141
Voegeli, Vincent (TranSystems Corporation/USA) 6(7):229
Vogt, Bob (Louisiana-Pacific Corporation/USA) 6(3):83
Volkering, Frank (TAUW bv/THE NETHERLANDS) 6(4):91
von Arb, Michelle (University of Iowa) 6(3):1
Vondracek, James E. (Ashland Inc./USA) 6(5):121
Vos, Johan (VITO/BELGIUM) 6(9):87
Voscott, Hoa T. (Camp Dresser & McKee, Inc./USA) 6(7):305
Vough, Lester R. (University of Maryland/USA) 6(5):77

Waisner, Scott A. (TA Environmental, Inc./USA) 6(4):59; 6(10):115

Walecka-Hutchison, Claudia M. (University of Arizona/USA) 6(9):231
Wall, Caroline (CEFAS Laboratory/UK) 6(10):337
Wallace, Steve (Lattice Property Holdings Plc./UK) 6(4):17
Wallis, F.M. (University of Natal/REP OF SOUTH AFRICA) 6(6):101; 6(9):79
Walton, Michelle R. (Idaho National Engineering & Environmental Laboratory/USA) 6(7):77
Walworth, James L. (University of Arizona/USA) 6(9):231
Wan, C.K. (Hong Kong Baptist University/CHINA) 6(6):73
Wang, Chuanyue (Rice University/USA) 6(5):85
Wang, Qingren (Chinese Academy of Sciences/CHINA [PRC]) 6(9):113
Wani, Altaf (Applied Research Associates, Inc./USA) 6(10):115
Wanty, Duane A. (The Gillette Company/USA) 6(7):87
Warburton, Joseph M. (Parsons Engineering Science/USA) 6(7):173
Watanabe, Masataka (National Institute for Environmental Studies/JAPAN) 6(5):321
Watson, James H.P. (University of Southampton/UK) 6(9):61
Wealthall, Gary P. (University of Sheffield/UK) 6(1):59
Weathers, Lenly J. (Tennessee Technological University/USA) 6(8):139
Weaver, Dallas E. (Scientific Hatcheries/USA) 6(1):91
Weaverling, Paul (Harding ESE/USA) 6(10):231
Weber, A. Scott (State University of New York at Buffalo/USA) 6(6):89
Weeber, Philip A. (Geotrans/USA) 6(10):163
Wendt-Potthoff, Katrin (UFZ Centre for Environmental Research/GERMANY) 6(9):43
Werner, Peter (Technical University of Dresden/GERMANY) 6(3):227; 6(8):105

West, Robert J. (The Dow Chemical Company/USA) 6(2):89
Westerberg, Karolina (Stockholm University/SWEDEN) 6(3):133
Weston, Alan F. (Conestoga-Rovers & Associates/USA) 6(1):99; 6(10):131
Westray, Mark (ThermoRetec Corp/USA) 6(7):1
Wheater, H.S. (Imperial College of Science and Technology/UK) 6(10):123
White, David C. (University of Tennessee/USA) 6(4):73; 6(5):305
White, Richard (EarthFax Engineering Inc/USA) 6(6):263
Whitmer, Jill M. (GeoSyntec Consultants/USA) 6(9):105
Wick, Lukas Y. (Swiss Federal Institute of Technology/SWITZERLAND) 6(3):251
Wickramanayake, Godage B. (Battelle/USA) 6(10):1
Widada, Jaka (The University of Tokyo/JAPAN) 6(4):51
Widdowson, Mark A. (Virginia Polytechnic Institute & State University/USA) 6(2):105; 6(5):1
Wieck, James M. (GZA GeoEnvironmental, Inc./USA) 6(7):165
Wiedemeier, Todd H. (Parsons Engineering Science, Inc./USA) 6(7):241
Wiessner, Arndt (UFZ - Centre for Environmental Research/GERMANY) 6(5):337
Wilken, Jon (Harding ESE/USA) 6(10):231
Williams, Lakesha (Southern University at New Orleans/USA) 6(5):145
Williamson, Travis (Battelle/USA) 6(10):245
Willis, Matthew B. (Cornell University/USA) 6(4):125
Willumsen, Pia Arentsen (National Environmental Research Institute/DENMARK) 6(3):141
Wilson, Barbara H. (Dynamac Corporation/USA) 6(1):129
Wilson, Gregory J. (University of Cincinnati/USA) 6(1):19

Wilson, John T. (U.S. EPA/USA) 6(1):43, 167
Wiseman, Lee (Camp Dresser & McKee Inc./USA) 6(7):133
Wisniewski, H.L. (Equilon Enterprises LLC/USA) 6(8):61
Witt, Michael E. (The Dow Chemical Company/USA) 6(2):89
Wong, Edwina K. (University of Guelph/CANADA) 6(6):185
Wong, J.W.C. (Hong Kong Baptist University/CHINA) 6(6):73
Wood, Thomas K. (University of Connecticut/USA) 6(5):199
Wrobel, John (U.S. Army/USA) 6(5):207

Xella, Claudio (Water & Soil Remediation S.r.l./ITALY) 6(6):179
Xing, Jian (Global Remediation Technologies, Inc./USA) 6(6):311

Yamamoto, Isao (Sumitomo Marine Research Institute/JAPAN) 6(10):345
Yamazaki, Fumio (Hyogo Prefectural Institute of Environmental Science/JAPAN) 6(5):321
Yang, Jeff (URS Corporation/USA) 6(2):239
Yerushalmi, Laleh (Biotechnology Research Institute/CANADA) 6(3):165
Yoon, Woong-Sang (Sam) (Battelle/USA) 6(7):13
Yoshida, Takako (The University of Tokyo/JAPAN) 6(4):51; 6(6):111
Yotsumoto, Mizuyo (Obayashi Corporation Ltd./JAPAN) 6(6):111
Young, Harold C. (Air Force Institute of Technology/USA) 6(2):173

Zagury, Gérald J. (École Polytechnique de Montréal/CANADA) 6(9): 27, 129
Zahiraleslamzadeh, Zahra (FMC Corporation/USA) 6(7):221
Zaluski, Marek H. (MSE Technology Applications/USA) 6(9):35
Zappi, Mark E. (Mississippi State University/USA) 6(9):241

Zelennikova, Olga (University of Connecticut/USA) *6*(7):69
Zhang, Chuanlun L. (University of Missouri/USA) *6*(9):165
Zhang, Wei (Cornell University/USA) *6*(4):125
Zhang, Zhong (University of Nevada Las Vegas/USA) *6*(9):257

Zheng, Zuoping (University of Oslo/NORWAY) *6*(2):181
Zocca, Chiara (Universita degli Studi di Verona/ITALY) *6*(3):267
Zwick, Thomas C. (Battelle/USA) *6*(10):1

KEYWORD INDEX

This index contains keyword terms assigned to the articles in the ten-volume proceedings of the Sixth International In Situ and On-Site Bioremediation Symposium (San Diego, California, June 4-7, 2001). Ordering information is provided on the back cover of this book.

In assigning the terms that appear in this index, no attempt was made to reference all subjects addressed. Instead, terms were assigned to each article to reflect the primary topics covered by that article. Authors' suggestions were taken into consideration and expanded or revised as necessary. The citations reference the ten volumes as follows:

6(1): Magar, V.S., J.T. Gibbs, K.T. O'Reilly, M.R. Hyman, and A. Leeson (Eds.), *Bioremediation of MTBE, Alcohols, and Ethers*. Battelle Press, Columbus, OH, 2001. 249 pp.

6(2): Leeson, A., M.E. Kelley, H.S. Rifai, and V.S. Magar (Eds.), *Natural Attenuation of Environmental Contaminants*. Battelle Press, Columbus, OH, 2001. 307 pp.

6(3): Magar, V.S., G. Johnson, S.K. Ong, and A. Leeson (Eds.), *Bioremediation of Energetics, Phenolics, and Polycyclic Aromatic Hydrocarbons*. Battelle Press, Columbus, OH, 2001. 313 pp.

6(4): Magar, V.S., T.M. Vogel, C.M. Aelion, and A. Leeson (Eds.), *Innovative Methods in Support of Bioremediation*. Battelle Press, Columbus, OH, 2001. 197 pp.

6(5): Leeson, A., E.A. Foote, M.K. Banks, and V.S. Magar (Eds.), *Phytoremediation, Wetlands, and Sediments*. Battelle Press, Columbus, OH, 2001. 383 pp.

6(6): Magar, V.S., F.M. von Fahnestock, and A. Leeson (Eds.), *Ex Situ Biological Treatment Technologies*. Battelle Press, Columbus, OH, 2001. 423 pp.

6(7): Magar, V.S., D.E. Fennell, J.J. Morse, B.C. Alleman, and A. Leeson (Eds.), *Anaerobic Degradation of Chlorinated Solvents*. Battelle Press, Columbus, OH, 2001. 387 pp.

6(8): Leeson, A., B.C. Alleman, P.J. Alvarez, and V.S. Magar (Eds.), *Bioaugmentation, Biobarriers, and Biogeochemistry*. Battelle Press, Columbus, OH, 2001. 255 pp.

6(9): Leeson, A., B.M. Peyton, J.L. Means, and V.S. Magar (Eds.), *Bioremediation of Inorganic Compounds*. Battelle Press, Columbus, OH, 2001. 377 pp.

6(10): Leeson, A., P.C. Johnson, R.E. Hinchee, L. Semprini, and V.S. Magar (Eds.), *In Situ Aeration and Aerobic Remediation*. Battelle Press, Columbus, OH, 2001. 391 pp.

A

abiotic/biotic dechlorination **6(8)**:193
acenaphthene **6(5)**:253
acetate as electron donor **6(3)**:51; **6(9)**:297
acetone **6(2)**:49
acid mine drainage, (*see also* mine tailings) **6(9)**:1, 9, 27, 35, 43, 53
acrylic vessel **6(5)**:321
actinomycetes **6(10)**:211
activated carbon biomass carrier **6(6)**:311; **6(8)**:113

activated carbon **6(8)**:105
adsorption **6(3)**:243; **6(5)**:253; **6(6)**:377; **6(7)**:77; **6(8)**:131; **6(9)**:86
advanced oxidation **6(1)**:121; **6(10)**:33
aerated submerged **6(10)**:329
aeration **6(6)**:203
anaerobic/aerobic treatment **6(6)**:361; **6(7)**:229
age dating **6(5)**:231, 237
air sparging **6(1)**:115, 175; **6(2)**:239; **6(9)**:215; **6(10)**:1, 9, 41, 49, 65, 101, 115, 123, 163, 223
alachlor **6(6)**:9
algae **6(5)**:181
alkaline phosphatase **6(9)**:165
alkane degradation **6(5)**:313
alkylaromatic compounds **6(6)**:173
alkylbenzene **6(2)**:19
alkylphenolethoxylate **6(5)**:215
Amaranthaceae **6(5)**:165
Ames test **6(6)**:249
ammonia **6(1)**:175; **6(5)**:337
amphipod toxicity test **6(5)**:321
anaerobic **6(1)**:35, 43; **6(3)**:91; 205; **6(5)**:17, 25, 261, 297, 313; **6(6)**:133; **6(7)**:249, 297; **6(9)**:147, 303
anaerobic biodegradation **6(1)**:137; **6(5)**:1; **6(8)**:167
anaerobic bioventing **6(3)**:9
anaerobic petroleum degradation **6(5)**:25
anaerobic sparging **6(7)**:297
aniline **6(6)**:149
Antarctica **6(2)**:57
anthracene **6(3)**:165, 251; **6(6)**:73
aquatic plants **6(5)**:181
arid-region soils **6(9)**:231
aromatic dyes **6(6)**:369
arsenic **6(2)**:239, 261; **6(5)**:173; **6(9)**: 97, 129
atrazine **6(5)**:181; **6(6)**:9
azoaromatic compounds **6(6)**:149
Azomonas **6(6)**:219

B

bacterial transport **6(8)**:1
barrier technologies **6(1)**:11; **6(3)**:165; **6(7)**:289; **6(8)**:61, 79, 87, 105, 121; **6(9)**:27, 71, 195, 209, 309
basidiomycete **6(6)**:101
benthic **6(10)**:337

benzene **6(1)**:1, 67, 75, 145, 167, 203; **6(4)**:91,117; **6(8)**:87; **6(10)**:123
benzene, toluene, ethylbenzene, and xylenes (BTEX) **6(1)**:43, 51, 59, 107, 129, 167, 195; **6(2)**:11, 19, 137, 215, 223, 270; **6(4)**:99; **6(5)**:33; **6(7)**:133; **6(8)**:105; **6(10)**: 1, 23, 49, 65, 95, 123, 131
benzo(a)pyrene **6(3)**:149; **6(6)**:101
benzo(e)pyrene **6(3)**:149
BER, *see* biofilm-electrode reactor
bioassays **6(3)**:219
bioaugmentation **6(1)**:11; **6(3)**:133; **6(4)**:59; **6(6)**:9, 43, 111; **6(7)**:125; **6(8)**:1, 11, 19, 27, 43, 53, 61, 147, 175
bioavailability **6(3)**:115, 157, 173, 189, 51; **6(4)**:7; **6(5)**:253, 279, 289; **6(6)**:1
bioavailable FeIII assay **6(8)**:209
biobarrier **6(1)**:11; **6(3)**:165; **6(7)**:289; **6(8)**:61, 79, 105, 121; **6(9)**:27, 71, 209, 309
BIOCHLOR model **6(2)**:155
biocide **6(7)**:321, 333
biodegradability **6(6)**:193
biodegradation **6(1)**:19,153; **6(3)**:165, 181, 205, 235; **6(10)**:187
biofilm **6(3)**:251; **6(4)**:149; **6(8)**:79; **6(9)**:201, 303
biofilm-electrode reactor (BER) **6(9)**:201
biofiltration **6(4)**:149
biofouling **6(7)**:321, 333
bioindicators **6(1)**:1; **6(3)**:173; **6(5)**:223
biological carbon regeneration **6(8)**:105
bioluminescence **6(1)**:1; **6(3)**:173; **6(4)**:45
biopile **6(6)**:81, 127, 141, 227, 249, 287
bioreactors **6(1)**:91; **6(6)**:361; **6(8)**:11, 35; **6(9)**:1, 265, 281, 303, 315; **6(10)**:171, 211
biorecovery of metals **6(9)**:9
bioreporters **6(4)**:45
biosensors **6(1)**:1
bioslurping **6(10)**:245, 253, 267, 275
bioslurry and bioslurry reactors **6(3)**:189; **6(6)**:51, 65
biosparging **6(10)**:115, 163
biostabilization **6(6)**:89
biostimulation **6(6)**:43
biosurfactant **6(3)**:243; **6(7)**:53
bioventing **6(10)**:109, 115, 131
biphasic reactor **6(3)**:181

biological oxygen demand (BOD) **6(10)**:311
BTEX, *see* benzene, toluene, ethylbenzene, and xylenes
Burkholderia cepacia **6(1)**:153; **6(7)**:117; **6(8)**:53
butane **6(1)**:137, 161
butyrate **6(7)**:289

C

cadmium **6(3)**:91; **6(9)**:79, 147
carAa, see carbazole 1,9a-dioxygenase gene
carbazole-degrading bacterium **6(6)**:111
carbazole 1,9a-dioxygenase gene (*carAa*) **6(4)**:51
Carbokalk **6(9)**:43
carbon isotope **6(4)**:91, 99, 109, 117; **6(10)**:115
carbon tetrachloride (CT) **6(2)**:81, 89; **6(5)**:113; **6(7)**:241; **6(8)**:185, 193
cesium-137 **6(5)**:231
CF, *see* chloroform
charged coupled device camera **6(2)**:207
chelators addition (EDGA, EDTA) **6(5)**:129, 137, 145, 151; **6(9)**:123, 147
chemical oxidation **6(7)**:45
chicken manure **6(9)**:289
chlorinated ethenes **6(7)**:27, 61, 69, 109; **6(10)**:163, 201, 231
chlorinated solvents **6(2)**:145; **6(7)**:all; **6(8)**:19; **6(10)**:231
chlorobenzene **6(8)**:105
chloroethane **6(2)**:113; **6(7)**:133, 249
chloroform (CF) **6(2)**:81; **6(8)**:193
chloromethanes **6(8)**:185
chlorophenol **6(3)**:75, 133
chlorophyll fluorescence **6(5)**:223
chromated copper arsenate **6(9)**:129
chromium (Cr[VI]) **6(8)**:139, 147; **6(9)**:129, 139, 315
chrysene **6(6)**:101
citrate and citric acid **6(5)**:137; **6(7)**:289
cleanup levels **6(6)**:1
coextraction method **6(4)**:51
Coke Facility waste **6(2)**:129
combined chemical toxicity (*see also* toxicity) **6(5)**:305
cometabolic air sparging **6(10)**:145, 155, 223

cometabolism **6(1)**:137, 145, 153, 161; **6(2)**:19; **6(6)**:81, 141; **6(7)**:117; **6(10)**:145, 155, 163, 171, 179, 193, 201, 211, 217, 223, 231; 239
competitive inhibition **6(2)**:19
composting **6(3)**:83; **6(5)**:129, **6(6)**:73, 119, 165, 257; **6(7)**:141
constructed wetlands **6(5)**:173, 329
contaminant aging **6(3)**:157, 197
contaminant transport **6(3)**:115
copper **6(9)**:79, 129
cosolvent effects **6(1)**:175, 195, 203, 243
cosolvent extraction **6(7)**:125
cost analyses and economics of environmental restoration **6(1)**:129; **6(4)**:17; **6(8)**:121; **6(9)**:331; **6(10)**:65, 211
Cr(VI), *see* chromium
creosote **6(3)**:259; **6(4)**:59; **6(5)**:1, 237, 329; **6(6)**:81, 101, 141, 295
cresols **6(10)**:123
crude oil **6(5)**:313; **6(6)**:193, 249; **6(10)**:329
CT, *see* carbon tetrachloride
cyanide **6(9)**:331
cytochrome P-450 **6(6)**:17

D

2,4-DAT, *see* diaminotoluene
DCA, *see* dichloroethane
1,1-DCA, *see* 1,1-dichloroethane
1,2-DCA, *see* 1,2-dichloroethane
DCE, *see* dichloroethene
1,1-DCE, *see* 1,1-dichloroethene
1,2-DCE, *see* 1,2-dichloroethene
c-DCE, *see* cis-dichloroethene
DCM, *see* dichloromethane
DDT, *see also* dioxins *and* pesticides **6(6)**:157
2,4-DNT, *see* dinitrotoluene
dechlorination kinetics **6(2)**:105; **6(7)**:61
dechlorination **6(2)**:231; **6(3)**:125; **6(5)**:95; **6(7)**:13, 61, 165, 173, 333; **6(8)**:19, 27, 43
DEE, *see* diethyl ether
Dehalococcoides ethenogenes **6(8)**:19, 43
dehalogenation **6(8)**:167
denaturing gradient gel electrophoresis (DGGE) **6(1)**:19; **6(4)**:35

denitrification **6(2)**:19; **6(4)**:149; **6(5)**:17, 261; **6(8)**:95; **6(9)**:179, 187, 195, 201, 209, 223, 309
dense, nonaqueous-phase liquid (DNAPL) **6(7)**:13, 19, 35, 181; **6(10)**:319
depletion rate **6(1)**:67
desorption **6(3)**:235, 243; **6(5)**:253; **6(6)**:377; **6(7)**:53, 77; **6(8)**:131
DGGE, *see* denaturing gradient gel electrophoresis
DHPA, *see* dihydroxyphenylacetate
dialysis sampler **6(5)**:207
diaminotoluene (2,4-DAT) **6(6)**:149
dibenzofuran-degrading bacterium **6(6)**:111
dibenzo-p-dioxin **6(6)**:111
dibenzothiophene **6(3)**:267
dichlorodiethyl ether **6(10)**:301
dichloroethane (DCA) **6(2)**:39; **6(7)**:289
1,1-dichloroethane (1,1-DCA; 1,2-DCA) **6(2)**:113; **6(5)**:207; **6(7)**:133, 165
1,2-dichloroethane (1,2-DCA) **6(5)**:207
dichloroethene, dichloroethylene **6(2)**:97, 155; **6(4)**:125; **6(5)**:105,113; **6(7)**:157, 197
cis-dichloroethene, *cis*-dichloroethylene (c-DCE) **6(2)**:39, 65, 73; 105, 173; **6(5)**:33, 95, 207; **6(7)**:1, 13, 61, 133, 141, 149, 165, 173, 181, 189, 205, 213, 221, 249, 273, 281, 289, 297, 305; **6(8)**:11, 19, 27, 43, 73, 105, 157, 209; **6(10)**:41, 145, 155, 179, 201
1,1-dichloroethene, 1,1-dichloroethylene (1,1-DCE) **6(2)**:39; **6(7)**:165, 229; **6(8)**:157; **6(10)**:231
1,2-dichloroethene and 1,2-dichloroethylene (1,2-DCE) **6(2)**:113
dichloromethane (DCM) **6(2)**:81; **6(8)**:185
diesel fuel **6(1)**:175; **6(2)**:57; **6(5)**:305; **6(6)**:81, 141, 165; **6(10)**:9
diesel-range organics (DRO) **6(10)**:9
diethyl ether (DEE) **6(1)**:19
dihydroxyphenylacetate (DHPA) **6(4)**:29
diisopropyl ether (DIPE) **6(1)**:19, 161
1,3-dinitro-5-nitroso-1,3,5-triazacyclohexane (MNX) (*see also* explosives *and* energetics) **6(3)**:51; **6(8)**:175
dinitrotoluene (2,4-DNT) **6(3)**:25, 59; **6(6)**:127, 149
dioxins **6(6)**:111

DIPE, *see* diisopropyl ether
dissolved oxygen **6(2)**:189, 207
16S rDNA sequencing **6(8)**:19
DNAPL, *see* dense, nonaqueous-phase liquid
DNX, *see* explosives and energetics
DRO, *see* diesel-range organics
dual porosity aquifer **6(1)**:59
dyes **6(6)**:369

E

ecological risk assessment **6(4)**:1
ecotoxicity, (*see also* toxicity) **6(1)**:1; **6(4)**:7
ethylenedibromide (EDB) **6(10)**:65
EDGA, *see* chelate addition
EDTA, *see* chelate addition
effluent **6(4)**:1
electrokinetics **6(9)**:241, 273
electron acceptors and electron acceptor processes **6(2)**:1, 137, 163, 231; **6(5)**:17, 25, 297; **6(7)**:19
electron donor amendment **6(3)**:25, 35, 51, 125; **6(7)**:69, 103,109, 141, 181, 249, 289, 297; **6(8)**:73; **6(9)**:297, 315
electron donor delivery **6(7)**:19, 27, 133, 173, 213, 221, 265, 273, 281, 305
electron donor mass balance **6(2)**:163
electron donor transport **6(4)**:125; **6(7)**:133; **6(9)**:241
embedded carrier **6(9)**:187
encapsulated bacteria **6(5)**:269
enhanced aeration **6(10)**:57
enhanced desorption **6(7)**:197
environmental stressors **6(4)**:1
enzyme induction **6(6)**:9; **6(10)**:211
ERIC sequences **6(4)**:29
ethane **6(2)**:113; **6(7)**:149
ethanol 6(1):19,167,175, 195, 203; **6(5)**:243; **6(6)**:133; **6(9)**:289
ethene and ethylene **6(2)**:105,113; **6(5)**:95; **6(7)**:1, 95, 133, 141, 205, 281, 297, 305; **6(8)**:11, 43, 167, 175, 209
ethylene dibromide **6(10)**:193
explosives and energetics **6(3)**:9, 17, 25, 35, 43, 51, 67; **6(5)**:69; **6(6)**:119, 127, 133; **6(7)**:125

F

fatty acids **6(5)**:41
Fe(II), *see* iron
Fenton's reagent **6(6)**:157
fertilizer **6(5)**:321; **6(6)**:35; **6(10)**:337
fixed-bed and fixed-film reactors **6(5)**:221, 337; **6(6)**:361; **6(9)**:303
flocculants **6(6)**:279
flow sensor **6(10)**:293
fluidized-bed reactor **6(1)**:91; **6(6)**:133, 311; **6(9)**:281
fluoranthene **6(3)**:141; **6(6)**:101
fluorogenic probes **6(4)**:51
food safety **6(9)**:113
formaldehyde **6(6)**:329
fractured shale **6(10)**:49
free-product recovery **6(6)**:211
Freon **6(2)**:49
fuel oil **6(5)**:321
fungal remediation **6(3)**:75, 99; **6(5)**:61, 279; **6(6)**:17, 101, 157, 263, 319, 329, 369
Funnel-and-Gate™ **6(8)**:95

G

gas flux **6(6)**:185
gasoline **6(1)**:35, 75, 161, 167, 195; **6(10)**:115
gasoline-range organics (GRO) **6(10)**:9
manufactured gas plants and gasworks **6(2)**:137; **6(10)**:123
GCW, *see* groundwater circulating well
gel-encapsulated biomass **6(8)**:35
GEM, see genetically engineered microorganisms
genetically engineered microorganisms (GEM) **6(4)**:45; **6(5)**:199; **6(7)**:125
genotoxicity, (*see also* toxicity) **6(3)**:227
Geobacter **6(3)**:1
geochemical characterization **6(4)**:91
geographic information system (GIS) **6(2)**:163
geologic heterogeneity **6(2)**:11
germination index 6(3):219; **6(6)**:73
GFP, *see* green fluorescent protein
GIS, *see* geographic information system
glutaric dialdehyde dehydrogenase **6(4)**:81
Gordonia terrae **6(1)**:153
green fluorescent protein (GFP) **6(5)**:199
GRO, see gasoline-range organics

groundwater **6(3)**:35; **6(8)**: 35, 87, 121; **6(10)**:231
groundwater circulating well (GCW) **6(7)**:229, 321; **6(10)**:283, 293

H

H_2 gas, *see* hydrogen
H_2S, *see* hydrogen sulfide
halogenated hydrocarbons **6(9)**:61
halorespiration **6(8)**:19
heavy metal **6(2)**:239; **6(5)**:137, 145, 157, 165, 173; **6(6)**:51; **6(9)**:53, 61, 71, 79, 86, 97, 113, 129, 147
herbicides **6(5)**:223; **6(6)**:35
hexachlorobenzene **6(3)**:99
hexane **6(3)**:181, **6(6)**:329
HMX, *see* explosives and energetics
hollow fiber membranes **6(5)**:269
hopane **6(6)**:193; **6(10)**:337
hornwort **6(5)**:181
HRC® (a proprietary hydrogen-release compound) **6(3)**:17, 25, 107; **6(7)**:27, 103, 157, 189, 197, 205, 221, 257, 305, **6(8)**:157, 209
^{2}H-tetradecane (*see also* tetradecane) **6(2)**:27
humates **6(1)**:99
hybrid treatment **6(10)**:311
hydraulic containment **6(8)**:79
hydraulically facilitated remediation **6(2)**:239
hydrocarbon **6(6)**:235; **6(10)**:329
hydrogen (H_2 gas) 6(2):199; **6(9)**:201
hydrogen injection, in situ **6(7)**:19
hydrogen isotope **6(4)**:91
hydrogen peroxide **6(1)**:121; **6(6)**:353; **6(10)**:33
hydrogen release compound, see HRC®
hydrogen sulfide (H_2S) **6(9)**:123
hydrogen **6(2)**:231, **6(7)**:61, 305
hydrolysis **6(1)**:83
hydrophobicity **6(3)**:141
hydroxyl radical **6(1)**:121
hydroxylamino TNT intermediates **6(5)**:85

I

immobilization **6(8)**:53
immobilized cells **6(8)**:121
immobilized soil bioreactor **6(10)**:171

in situ oxidation **6(7)**:1
industrial effluents **6(6)**:303, 361
inhibition **6(9)**:17
injection strategies, in situ **6(7)**:19, 133, 173, 213, 221, 265, 273, 305, 313; **6(9)**:223; **6(10)**:23, 163
insecticides **6(6)**:27
intrinsic biodegradation **6(2)**:89, 121
intrinsic remediation, *see* natural attenuation
ion migration **6(9)**:241
iron (Fe[II]) **6(5)**:1
iron barrier **6(8)**:139, 147, 157, 167
iron oxide **6(3)**:1
iron precipitation **6(3)**:211
iron-reducing processes **6(2)**:121; **6(3)**:1; **6(5)**:1, 17, 25; **6(6)**:149; **6(8)**:193, 201, 209; **6(9)**:43, 323
IR-spectroscopy **6(4)**:67
isotope analyses **6(2)**:27; **6(4)**:91; **6(8)**:27
isotope fractionation **6(4)**:99, 109, 117

J
jet fuel **6(10)**:95, 139

K
KB-1 strain **6(8)**:27
kerosene **6(6)**:219
kinetics **6(8)**:131, **6(1)**:1, 19, 27, 167; **6(2)**:11, 19, 105; **6(3)**:173; **6(4)**:131; **6(7)**:61
Klebsiella oxytoca **6(7)**:117
Kuwait **6(6)**:249

L
laccase **6(3)**:75; **6(6)**:319
lactate and lactic acid **6(7)**:103, 109, 165, 181, 213, 265, 281, 289; **6(8)**:139; **6(9)**:155, 273
lagoons **6(6)**:303
land treatment units (LTU) **6(6)**:1; **6(6)**:81, 141, 287, 295
landfarming **6(3)**:259; **6(4)**:59; **6(5)**:53, 279; **6(6)**:1, 43, 59, 179, 203, 211, 235
landfills **6(2)**:145, 247; **6(4)**:91; **6(8)**:113
leaching **6(9)**:187
lead **6(5)**:129, 145, 151, 157

lead-210 **6(5)**:231
light, nonaqueous-phase liquids (LNAPL) **6(1)**:59; **6(4)**:35; **6(10)**:57, 109, 245, 253, 275
lindane, (*see also* pesticides) **6(5)**:189
linuron (*see also* herbicides) **6(5)**:223
LNAPL, *see* light, nonaqueous-phase liquids
Lolium multiflorum **6(5)**:9
LTU, *see* land treatment units
lubricating oil **6(6)**:173
luciferase **6(3)**:133
lux **6(4)**:45

M
mackinawite **6(9)**:155
macrofauna **6(10)**:337
magnetic separation **6(9)**:61
magnetite **6(3)**:1; **6(8)**:193
manganese **6(2)**:261
manufactured gas plant (MGP) **6(2)**:19; **6(3)**:211, 227; **6(10)**:123
mass balance **6(2)**:163
mass transfer limitation **6(3)**:157
mass transfer **6(1)**:67
MC-100, see mixed culture
media development **6(9)**:147
Meiofauna **6(5)**:305; **6(10)**:337
membrane **6(5)**:269; **6(9)**:1, 265
metabolites **6(3)**:227
metal reduction **6(8)**:1
metal precipitation **6(9)**:9, 165
metals, biorecovery of **6(9)**:9
metals speciation **6(9)**:129
metal toxicity (*see also* toxicity) **6(9)**:17, 129
metals **6(5)**:129, 305; **6(8)**:1; **6(9)**:9, 17, 27, 105, 123, 129, 155, 165
methane oxidation **6(10)**:171, 187, 193, 201, 223, 231
methane **6(1)**:183; **6(8)**:113
methanogenesis **6(1)**:35, 43, 183; **6(3)**:205; **6(9)**:147
methanogens **6(3)**:91
methanol **6(1)**:183; **6(7)**:141, 289, 297
methanotrophs **6(10)**:171, 187, 201
methylene chloride **6(2)**:39; **6(10)**:231
Methylosinus trichosporium **6(10)**:187
methyl *tert*-butyl ether *or* methyl *tertiary*-butyl ether (MTBE) **6(1)**:1, 11, 19, 27, 35, 43, 51, 59, 67, 75, 83, 91, 107,

115, 121, 129, 137, 145, 153,161, 195, **6(2)**:215; **6(8)**:61; **6(10)**:1, 65
MGP, *see* manufactured gas plant
microbial heterogeneity **6(4)**:73
microbial isolation **6(3)**:267
microbial population dynamics **6(4)**:35
microbial regrowth **6(2)**:253; **6(7)**:1, 13; **6(10)**:319
microcosm studies **6(7)**:109; **6(10)**:179
microencapsulation **6(8)**:53
microfiltration **6(9)**:201
microporous membrane **6(9)**:265
microtox assay **6(3)**:227
mine tailings (*see also* acid mine drainage) **6(5)**:173; **6(9)**:27, 71
mineral oil **6(5)**:279, 289; **6(6)**:59
mineralization **6(2)**:121; **6(3)**:165; **6(6)**:165; **6(8)**:175; **6(9)**:139, 155
MIP, *see* membrane interface probe
mixed culture **6(8)**:61
mixed wastes **6(3)**:91; **6(7)**:133; **6(9)**:139
MNX, *see* 1,3-dinitro-5-nitroso-1,3,5-triazacyclohexane
modeling **6(1)**:51; **6(2)**:105, 155, 181, **6(4)**:125, 131, 139, 149; **6(6)**:339, 377; **6(8)**:185; **6(9)**:27, 105; **6(10)**:163
moisture content **6(2)**:247
molasses as electron donor **6(3)**:35; **6(7)**:53, 103, 149, 173; **6(9)**:315
monitored natural attenuation (*see also* natural attenuation) **6(1)**:183, **6(2)**:11, 163, 199, 223, 253, 261
monitoring techniques **6(2)**:27,189, 199, 207; **6(4)**:59
motor oil **6(5)**:53
MPE, *see* multiphase extraction
multiphase extraction (MPE) well design **6(10)**:245, 259
MTBE, *see* methyl *tert*-butyl ether
multiphase extraction **6(10)**:245, 253, 259, 267, 275
municipal solid waste **6(2)**:247
Mycobacterium sp. IFP 2012 **6(1)**:153
Mycobacterium adhesion **6(3)**:251
mycoremediation **6(6)**:263

N

naphthalene **6(1)**:1; **6(2)**:121; **6(3)**:173, 227; **6(5)**:1, 253; **6(6)**:51; **6(8)**:95, **6(9)**:139; **6(10)**:123

NAPL, *see* nonaqueous-phase liquid
natural attenuation **6(1)**:27, 35, 43, 51, 59, 75, 83, 183, 195; **6(2)**:1,39, 73, 81, 89, 97, 105, 137, 145, 173, 181, 215; **6(4)**:91, 99, 117; **6(5)**:33, 189, 321; **6(8)**:185, 209; **6(9)**:179; **6(10)**:115, 163
natural gas **6(10)**:193
natural organic carbon **6(2)**:261
natural organic matter **6(2)**:81, 97; **6(8)**:201
natural recovery **6(5)**:132, 231
nitrate contamination **6(9)**:173
nitrate reduction **6(3)**:51; **6(5)**:25; **6(9)**:331
nitrate utilization efficiency **6(6)**:353
nitrate **6(2)**:1; **6(3)**:17, 43; **6(6)**:353; **6(8)**:95, 147; **6(9)**:179, 187, 195, 209, 223, 257
nitrification **6(4)**:149; **6(5)**:337; **6(9)**:215
nitroaromatic compounds (*see also* explosives and energetics) **6(3)**:59, 67; **6(6)**:149
nitrobenzene, *see also* explosives and energetics **6(6)**:149
nitrocellulose, *see also* explosives and energetics **6(6)**:119
nitrogen fixation **6(6)**:219
nitrogen utilization **6(9)**:231
nitrogenase **6(6)**:219
nitroglycerin, *see also* explosives and energetics **6(5)**:69
nitrotoluenes, *see also* explosives and energetics **6(6)**:127
nitrous oxide **6(8)**:113
^{13}C-NMR, *see* nuclear magnetic resonance spectroscopy
nonaqueous-phase liquids (NAPLs) **6(1)**:67, 203; **6(3)**:141; **6(7)**:249
nonylphenolethoxylates **6(5)**:215
nuclear magnetic resonance spectroscopy (^{13}C-NMR) **6(4)**:67
nutrient augmentation **6(3)**:59; **6(5)**:329; **6(6)**:257; **6(7)**:313; **6(9)**:331; **6(10)**:23
nutrient injection **6(10)**:101
nutrient transport **6(9)**:241

O

oily waste **6(4)**:35; **6(6)**:257; **6(10)**:337, 345
oil-coated stones **6(10)**:329

optimization *6(5)*:279
ORC® (a proprietary oxygen-release compound) *6(1)*:99,107; *6(2)*:215; *6(3)*:107; *6(7)*:229; *6(10)*:9, 15, 87, 95, 139
organic acids *6(2)*:39
organophosphorus *6(6)*:17, 27
advanced oxidation *6(6)*:157, *6(10)*:311
oxygen-release compound, *see* ORC®
oxygen-release material *6(10)*:73
oxygen respiration *6(9)*:231; *6(10)*:57
oxygenation *6(1)*:107, 145
ozonation *6(1)*:121; *6(10)*:33, 149, 301

P

packed-bed reactors *6(9)*:249; *6(10)*:329
PAHs, *see* polycyclic aromatic hydrocarbons
paper mill waste *6(4)*:1
paraffins *6(3)*:141
partitioning *6(9)*:129
PCBs, *see* polychlorinated biphenyls
PCP toxicity (*see also* toxicity) *6(3)*:125
PCP, *see* pentachlorophenol
PCR analysis, *see* polymerase chain reaction
pentachlorophenol (PCP) *6(3)*:83, 91, 99, 107, 115, 125; *6(5)*:329; *6(6)*:279, 287, 295, 329
percarbonate *6(10)*:73
perchlorate *6(9)*:249, 257, 265, 273, 281, 289, 297, 303, 309, 315
perchloroethene, perchloroethylene *6(7)*:53
permeable reactive barriers *6(3)*:1; *6(8)*: 73, 87, 95, 121, 139, 147, 157, 167, 175, 185; *6(9)*:71, 309, 323; *6(10)*:95
pesticides *6(5)*:189; *6(6)*:9, 17, 35
PETN reductase *6(5)*:69
petroleum hydrocarbon degradation *6(4)*:7; *6(5)*:9, 17, 25; *6(8)*:131; *6(10)*: 65, 101, 245, 345
phenanthrene *6(2)*:121; *6(3)*:227, 235, 243; *6(6)*:51, 65, 73
phenol *6(6)*:303, 319, 329
phenolic waste *6(6)*:311
phenol-oxidizing cultures *6(10)*:211, 217, 239
phenyldodecane *6(2)*:27
phosphate precipitation *6(9)*:165
PHOSter *6(10)*:65

photocatalysis *6(10)*:311
physical/chemical pretreatment *6(1)*:1, 51; *6(2)*:253; *6(3)*:149; *6(5)*:9, 33, 41, 53, 61, 69, 77, 85,105, 113, 121, 129,137, 145, 151, 157, 165, 189, 199, 207, 279, 337; *6(6)*:59, 157, 241; *6(7)*:1, 13; *6(9)*:113, 173; *6(10)*:239, 311, 319
phytotoxicity (*see also* toxicity) *6(5)*:41, 223
phytotransformation *6(5)*:85
pile-turner *6(6)*:249
PLFA, *see* phospholipid fatty acid analysis
polychlorinated biphenyls (PCBs) *6(2)*:39,105,173; *6(5)*:33, 61, 95, 113, 231, 289; *6(6)*:89, *6(7)*:13, 61, 69, 95, 109, 125, 133, 141, 149, 165, 181, 189, 197, 205, 213, 241, 249, 273, 297, 305; *6(8)*:11,19, 27, 43, 157, 167, 193, 209; *6(10)*:33, 41, 231, 283
polycyclic aromatic hydrocarbons (PAHs) *6(2)*:19, 121, 129, 137; *6(3)*:141, 149, 157, 165, 173, 181, 189, 197, 205, 211, 219, 227, 235, 243; *6(4)*:35, 45, 59, 67; *6(5)*:1, 9, 17, 41, 237, 243, 251, 253, 261, 269, 279, 289, 305, 329; *6(6)*:43, 51, 59, 65, 73, 81, 89, 101, 279, 295, 297; *6(7)*:125; *6(8)*:95; *6(9)*:139; *6(10)*:33, 123
polymerase chain reaction (PCR) analysis *6(4)*:29, 35, 51; *6(8)*:43
polynuclear aromatic hydrocarbons, *see* polycyclic aromatic hydrocarbons
poplar lipid fatty acid analysis (PLFA) *6(3)*:189
poplar trees *6(5)*:113, 121, 189
potassium permanganate *6(2)*:253; *6(7)*:1
precipitation *6(9)*:105; *6(10)*:301
pressurized-bed reactor *6(6)*:311
propane utilization *6(1)*:137; *6(10)*:145, 155, 179, 193
propionate *6(7)*:265, 289
Pseudomonas fluorescens *6(3)*:173
pyrene *6(3)*:165, 235; *6(4)*:67; *6(6)*: 65, 73, 101
pyridine *6(4)*:81

Keyword Index

R

RABITT, *see* reductive anaerobic biological in situ treatment technology
radium *6*(5):173
rapeseed oil *6*(6):65
RDX, *see* research development explosive
rebound *6*(10):1
recirculation well *6*(7):333, 341; *6*(10):283
redox measurement and control *6*(1):35; *6*(2):11, 231; *6*(5):1; *6*(9):53
reductive anaerobic biological in situ treatment technology (RABITT) *6*(7):109
reductive dechlorination *6*(2):39, 65, 97, 105, 145, 173; *6*(4):125; *6*(7):45, 53, 87, 103, 109, 133, 141, 149, 157,181, 197, 205, 213, 221, 249, 257, 265, 273, 289, 297; *6*(8):11, 73, 105, 157, 209
reductive dehalogenation *6*(7):69
reed canary grass *6*(5):181
research development explosive (RDX) *6*(3):1, 9, 17, 25, 35, 43, 51; *6*(6):133; *6*(8):175
respiration and respiration rates *6*(2):129; *6*(4):59; *6*(6):185, 227
respirometry *6*(6):127; *6*(10):217
rhizoremediation *6*(5):9, 61, 199
Rhodococcus opacus *6*(4):81
risk assessment *6*(2):215; *6*(4):1
16S rRNA sequencing *6*(8):43; *6*(9):147
rock-bed biofiltration *6*(4):149
rotating biological contactor *6*(9):79
rototiller *6*(6):203
RT3D *6*(10):163

S

salinity *6*(9):257
salt marsh *6*(5):313
SC-100, *see* single culture
Sea of Japan *6*(5):321
sediments *6*(3):91; *6*(5):231, 237, 253, 261, 269, 279, 289, 297, 305; *6*(6):51, 59; *6*(9):61
selenium *6*(9):323, 331
semivolatile organic carbon (SVOC) *6*(2):113
sheep dip *6*(6):27
Shewanella putrefaciens *6*(8):201
silicon oil *6*(3):141, 181
single culture *6*(8):61
site characterization *6*(10):139
site closure *6*(2):215
slow-release fertilizer *6*(2):57
sodium glycine *6*(9):273
soil treatment *6*(3):181; *6*(6):1
soil washing *6*(5):243; *6*(6):241
soil-vapor extraction (SVE) *6*(1):183; *6*(10):1, 41, 131, 223
solids residence time *6*(10):211
sorption *6*(5):215, 253; *6*(6):377; *6*(8):131; *6*(9):79, 105
source zone *6*(7):13, 19, 27, 181; *6*(10):267
soybean oil *6*(7):213
sparging *6*(10):33, 145, 155
stabilization *6*(6):89
substrate delivery *6*(7):281
sulfate reduction *6*(1):35; *6*(3):43, 91; *6*(5):261, 313; *6*(6):339; *6*(7):69, 95; *6*(8):139, 147, 193; *6*(9):1, 9, 17, 27, 35, 43, 61, 71, 86, 105, 123, 147
sulfide precipitation *6*(9):123
surfactants *6*(5):215; *6*(6):73; *6*(7):213, 321, 333; *6*(8):131
sustainability *6*(6):1
SVE, *see* soil vapor extraction
SVOC, *see* semivolatile organic carbon
synthetic pyrethroid *6*(6):27

T

TCA, see trichlorethane
1,1,1-TCA, *see* 1,1,1-trichloroethane
1,1,2-TCA, *see* 1,1,2-trichloroethane
2,4,6-TCP, *see* 2,4,6-trichlorophenol
1,1,1,2-TeCA,*see* tetrachloroethane
1,1,2,2-TeCA, *see* tetrachloroethane
1,3,5-TNB, *see* 1,3,5-trinitrobenzene
TAME, *see* tertiary methyl-amyl ether
TBA, *see* tertiary butyl alcohol
TBF, *see* tertiary butyl formate
TCE oxidation, *see* trichloroethene, trichloroethylene
TCE, *see* trichloroethene
TCP, *see* trichlorophenol
t-DCE, *see* trans-dichloroethene, trans-dichloroethylene
technology comparisons *6*(7):45; *6*(9):323
terrazyme *6*(10):345

tertiary butyl alcohol (TBA) **6(1)**:19, 27, 35, 51, 59, 91, 145, 153, 161
tertiary butyl formate (TBF) **6(1)**:145, 161
tertiary methyl-amyl ether (TAME) **6(1)**:59, 161
tetrachloroethane (1,1,1,2-TeCA, 1,1,2,2-TeCA) **6(5)**:207; **6(7)**:321, 341; **6(8)**:193
tetradecane (see also ^{2}H-tetradecane) **6(3)**:181
thermal desorption **6(3)**:189, **6(6)**:35
TNB, *see* trinitrobenzene
TNT, see trinitrotoluene
TNX, *see* 1,3,5-trinitroso-1,3,5-triazacyclohexane
tobacco plant **6(5)**:69
toluene **6(1)**:145; **6(2)**:181; **6(7)**:95; **6(8)**:35, 131
total petroleum hydrocarbons (TPH) **6(2)**:1; **6(5)**:9; **6(6)**:127, 173, 179, 193, 227, 241, 249; **6(10)**:15, 73, 115, 337
toxicity **6(1)**:1; **6(3)**:67, 189, 227; **6(4)**:7; **6(5)**:41, 61, 223, 305; **6(9)**:17, 129
TPH, *see* total petroleum hydrocarbons
trace gas emissions **6(6)**:185
trans-dichloroethene, trans-dichloroethylene **6(5)**:95, 207; **6(7)**:165
transgenic plants **6(5)**:69
transpiration **6(5)**:189
Trecate oil spill **6(6)**:241; **6(10)**:109
trichloroethane (TCA) **6(7)**:241, 281
1,1,1-trichloroethane (1,1,1-TCA; 1,1,2-TCA) **6(2)**:39, 113, 464; **6(5)**:207; **6(7)**:87,165, 281
1,1,2-trichloroethane (1,1,2-TCA) **6(5)**:207
trichloroethene, trichloroethylene (TCE) **6(2)**:39, 65, 73, 97, 105, 113, 155, 173, 253; **6(4)**:125; **6(5)**:33, 95, 105, 113, 207; **6(7)**:1, 13, 53, 61, 69, 77, 87, 109, 117, 133, 141, 149, 157, 181, 189, 197, 205, 213, 221, 241, 249, 265, 273, 281, 297, 305, **6(8)**:11, 19, 27, 35, 43, 53, 73, 105,147, 157, 193, 209; **6(10)**:41, 131, 145, 155, 163, 171, 179, 187, 201, 211, 217, 223, 231, 239, 283, 319
2,4,6-trichlorophenol (2,4,6-TCP) **6(3)**:75; **6(8)**:121
trichlorotrifluoroethane **6(2)**:49

trinitrobenzene (TNB) **6(3)**:9, 25
1,3,5-trinitroso-1,3,5-triazacyclohexane (TNX) **6(8)**:175
trinitrotoluene (TNT) **6(3)**:35, 67; **6(5)**:69, 77, 85; **6(6)**:133

U

underground storage tank (UST) **6(1)**:67, 129
uranium **6(5)**:173; **6(7)**:77; **6(9)**:155, 165
UST, *see* underground storage tank

V

vacuum extraction **6(1)**:115
vadose zone **6(1)**:183; **6(2)**:39, 65, 97, 105, 113, 155, 173; **6(3)**:9; **6(5)**:33, 105; **6(7)**:1,13, 61, 133, 141, 197, 205, 213, 249, 273, 281, 305; **6(8)**:11,19, 43, 73, 157, 209; **6(10)**:41, 163
vegetable oil **6(6)**:65; **6(7)**103, 213, 241, 249
vinyl chloride **6(2)**:73; **6(4)**:109; **6(5)**:95; **6(7)**:95,149, 157, 165, 173, 289, 297; **6(10)**:231
vitamin B$_{12}$ **6(7)**:321, 333, 341
VOCs, *see* volatile organic carbons
volatile fatty acid **6(7)**:61
volatile organic carbons (VOCs) **6(2)**:113, 189; **6(5)**:113, 121

W

wastewater treatment **6(5)**:215; **6(6)**:149; **6(9)**:173
water potential **6(9)**:231
weathering **6(4)**:7
wetlands **6(5)**:33, 95, 105, 313, 329; **6(9)**:97
white rot fungi, (*see also* fungal remediation) **6(3)**:75, 99; **6(6)**:17, 157, 263
windrow **6(6)**:81, 119, 141
wood preservatives **6(3)**:83, 259; **6(4)**:59; **6(6)**:279

X

xylene **6(1)**:67

Y
yeast extract *6(7)*:181

Z
zero-valent iron *6(8)*:157, 167; *6(9)*:71
zinc *6(4)*:91; *6(9)*:79